W0246218

Lecture Notes
in Control and Information Sciences   248

Editor: M. Thoma

Springer-Verlag London Ltd.

Yangquan Chen and Changyun Wen

Iterative Learning Control

Convergence, Robustness and Applications

With 69 Figures

 Springer

Series Advisory Board

A. Bensoussan · M.J. Grimble · P. Kokotovic · A.B. Kurzhanski · H. Kwakernaak · J.L. Massey · M. Morari

Editors

Yangquan Chen, PhD
Servo Group/Research & Development, Seagate Technology International, 63 The Fleming, Science Park Drive, Singapore Science Park, Singapore 118249

Changyun Wen, PhD
Block S2, School of EEE, Nanyang Technological University, Nanyang Avenue, Singapore 639798

British Library Cataloguing in Publication Data
Chen, Yangquan
   Iterative learning control : convergence, robustness and
   applications. - (Lecture notes in control and information
   sciences ; 248)
   1.Control theory 2.Intelligent control systems
   I.Title II. Wen, Changyun
   629.8
   ISBN 978-1-85233-190-0   ISBN 978-1-84628-539-4 (eBook)
   DOI 10.1007/978-1-84628-539-4
Library of Congress Cataloging-in-Publication Data
A catalog record for this book is available from the Library of Congress

Apart from any fair dealing for the purposes of research or private study, or criticism or review, as permitted under the Copyright, Designs and Patents Act 1988, this publication may only be reproduced, stored or transmitted, in any form or by any means, with the prior permission in writing of the publishers, or in the case of reprographic reproduction in accordance with the terms of licences issued by the Copyright Licensing Agency. Enquiries concerning reproduction outside those terms should be sent to the publishers.

© Springer-Verlag London 1999
Originally published by Springer-Verlag London Limited in 1999

The use of registered names, trademarks, etc. in this publication does not imply, even in the absence of a specific statement, that such names are exempt from the relevant laws and regulations and therefore free for general use.

The publisher makes no representation, express or implied, with regard to the accuracy of the information contained in this book and cannot accept any legal responsibility or liability for any errors or omissions that may be made.

Typesetting: Camera ready by authors

69/3830-543210 Printed on acid-free paper SPIN 10733087

This work is dedicated to

Our Parents, and

Huifang Dou and Duyun Chen (Jun Jun)
— Yangquan Chen

Xiu Zhou, Wen Wen, Wendy Wen and Qingyun Wen
— Changyun Wen

Preface

A SYSTEM is called a 'repetitive system' when it performs a given task repeatedly. Robotic and functional neuromuscular stimulation systems are examples of such systems. Iterative Learning Control (ILC), which is a relatively new area in the field of control, is proposed to control these types of systems.

Generally speaking, a system under control may have uncertainties in its dynamic model and its environment. One attractive point in ILC is the utilisation of the system repetitiveness to reduce such uncertainties and in turn to improve the control performance by operating the system repeatedly. ILC is a feedforward control strategy that updates, through iterative learning, control signals at every repeated operation. As the number of iterations increases, the system tracking error over the entire operation time period including the transient portion will decrease and eventually vanish. This may not be possible for conventional non-iterative learning control.

This book provides readers with a comprehensive coverage of iterative learning control and emphasises both theoretical and practical aspects. It provides some recent developments in ILC convergence and robustness analysis as well as issues in the ILC design. These include: *High-order Updating Laws, Discrete-time Domain Analysis, Use of the Current Iteration Tracking Error, Initial State Learning, Terminal Iterative Learning Control, ILC Design via Noncausal Filtering or Local Symmetrical Integral (LSI), Iterative Learning Identification* and so on. Several practical applications are included to illustrate the effectiveness of ILC. These applications include: *Rapid Thermal Processing Chemical Vapor Deposition (RTPCVD) Thickness Control in Wafer Fab Industry, Identification of Aerodynamic Drag Coefficient Curve* and *Trajectory Control in Functional Neuromuscular Stimulation (FNS) Systems.* The application examples provided are particularly useful to readers who wish to capitalise the system repetitiveness to improve system control performance.

The book can be used as a reference or a text for a course at graduate level. It is also suitable for self-study and for industry-oriented courses of continuing education. The knowledge background for this monograph would be some undergraduate and graduate courses including calculus, optimization and nonlinear system control. Some knowledge on optimal control and iteration theory would certainly be helpful. There are 12 chapters in this monograph. Chapter 1 is to give an introduction of ILC with an extended yet brief literature review. The next eight chapters, i.e. Chapters 2 - 9, provide some new theoretical developments in ILC, followed by two chapters on ILC applications. Conclusions are drawn and some recommendations for future research are given in the last chapter. However, the material in each chapter is largely independent so that the chapters may be used in almost any order desired.

The authors are grateful to Prof. Soh Yeng Chai, Head of the Control and Instrumentation Division, School of Electronic and Electrical Engineering, Nanyang Technological University (NTU), for his advice and encouragement. The first author is deeply thankful to Prof. Jian-Xin Xu and Prof. Tong Heng Lee, Dept. of Electrical Engineering, National University of Singapore (NUS). In addition, the first author would like to thank Mingxuan Sun, Dr. Huifang Dou, Prof. Kok Kiong Tan, Prof. Shigehiko Yamamoto for their collaboration and help. In particular, the first author appreciates Prof. Zenn Z. Bien, Prof. Kevin L. Moore and all the researchers in the ILC Web-link such as Prof. Minh Q. Phan, Prof. Theo J.A. de Vries, Prof. R. W. Longman, Prof. Dr.-Ing. M. Pandit, to name a few, for their constant encouragement during his Ph.D pursue. The authors would like to thank editorial and production staff at Springer-Verlag, especially Alison Jackson, Editorial Assistant (Engineering) and Nicholas Pinfield, Engineering Editor, Springer-Verlag London. Suggestions from Prof. Manfred Thoma, the LNCIS Series Editor, are also appreciated.

Last but not least, Yangquan Chen would like to thank his wife and son, Huifang Dou and Duyun Chen, Changyun Wen would like to thank his wife Xiu Zhou and his daughters Wen Wen, Wendy Wen and Qingyun Wen for their patience, understanding and complete support throughout this work.

Singapore, July 1999.   *Yangquan Chen and Changyun Wen*

Contents

1. Introduction

1.1 Background

1.1.1 Repetitive Systems

A 'REPETITIVE SYSTEM' is so called when the system performs a given task repeatedly. This kind of systems, also called the 'multi-pass' processes, was first investigated by J. B. Edwards and D. H. Owens [79, 80]. According to [79, 80], a repetitive dynamic system or multi-pass process is a process possessing the following two novel properties:

- *Repetitive action* in which the process repeats an operation by passing an object or information sequentially through the same dynamic machine or process; and
- *Interaction* between the state or output functions generated during successive operations.

Each individual cycle of operation is termed as a 'pass' or an iteration through the system dynamics. Typical examples of multi-pass processes include automatic ploughing, multi-machine systems such as vehicle conveyers, machining of metals and automatic coal cutting. The early research is to explore the new system characteristics and the main interests are in the stability analysis [79, 183, 207]. The interpretation of multi-pass systems in a framework of the 2-D system theory was found in [30, 204].

It is interesting to note some of the research suggestions from [80]:

- The inter-pass control policy clearly allows some potential for optimizing the behavior of successive batches as the pass number n increases [80, page 279].
- Again, of course, feedforward control based on previous-pass histories must be augmented by present-pass control [80, page 280].

However, there is an important feature behind such kind of system which was not emphasized in [80, 207]. That is, the system repetition can be utilized to improve the system performance by revising the controls from pass to pass. Later in [203, 202, 205, 208, 8, 209], this was realized.

1.1.2 Iterative Learning Control

The term 'iterative learning control' (ILC) was coined by Arimoto and his associates [17] for a better control of repetitive systems. In line with the discussion of the repetitive systems in Sec. 1.1.1, we can intuitively find that *learning is a bridge between knowledge and experience.* That is to say, the lack of knowledge is bridged by the experience. *Knowledge* and *Experience* in technical language can be obtained by 'modeling' and by 'repetitions after applying some *learning* control laws'. One way to describe the *learning* is as a process where the objective of achieving a desired result is obtained by experience when only partial knowledge about the plant is available.

Roughly speaking, the purpose of introducing the ILC is to utilize the system repetitions as *experience* to improve the system control performance even under incomplete *knowledge* of the system to be controlled. Based on Arimoto's formulation, the mathematical description of ILC is as follows.

- *Dynamic system and control task:*
  A general nonlinear dynamic system is considered. The system is controlled to track a given desired output $y_d(t)$ over a fixed time interval. The system is operated repeatedly and the state equation at the k-th repetition is described as follows:

$$\begin{cases} \dot{x}_k(t) = f(x_k(t), u_k(t)) \\ y_k(t) = g(x_k(t), u_k(t)) \end{cases} \tag{1.1}$$

  where $t \in [0, T]$; $x_k(t), y_k(t)$ and $u_k(t)$ are state, output and control variables respectively. Only the output $y_k(t)$ is assumed to be measureable and the tracking error at the k-th iteration is denoted by $e_k(t) \triangleq y_d(t) - y_k(t)$.

- *Postulates:*
  - **P1.** Every trial (pass, cycle, batch, iteration, repetition) ends in a fixed time of duration $T > 0$.
  - **P2.** A desired output $y_d(t)$ is given *a priori* over $[0, T]$.
  - **P3.** Repetition of the initial setting is satisfied, that is, the initial state $x_k(0)$ of the objective system can be set the same at the beginning of each iteration: $x_k(0) = x^0$, for $k = 1, 2, \cdots$.
  - **P4.** Invariance of the system dynamics is ensured throughout these repeated iterations.
  - **P5.** Every output $y_k(t)$ can be measured and therefore the tracking error signal, $e_k(t) = y_d(t) - y_k(t)$, can be utilized in the construction of the next input $u_{k+1}(t)$.
  - **P6.** The system dynamics are invertible, that is, for a given desired output $y_d(t)$ with a piecewise continuous derivative, there exists a unique input $u_d(t)$ that drives the system to produce the output $y_d(t)$.
- *Controller design task:* The controller design task is to find a recursive control law

$$u_{k+1}(t) = F(u_k(t), e_k(t)) \tag{1.2}$$

such that $e_k(t)$ vanishes as k tends to infinity.

The simpler the recursive form $F(\cdot, \cdot)$, the better it is for practical implementation of the iterative learning control law, as long as the convergence is assured and the convergence speed is satisfactory. The above set of postulates reflects the program learning and generation for the acquisition of various kinds of fast but skilled movements. Physiology suggests that ideal or desired pattern of motion must be acquired through a succession of trainings. Take the sports for example, once an idealized form of motion is pointed out by the coach, one must repeat the physical exercises to make his or her motion approach or converge to the ideal form. Through a sufficient number of trials or repetitions, a program is formed in the Central Nerve System (CNS) which can generate a succession of input command signals that excites a certain system of muscles and tendons related to that ideal motion pattern and realizes the desired motion form.

In the control community, a new research field has thus been developed to formulate controllers that achieve zero tracking error over the whole operational time period of the process including the transient response in the presence of disturbances with a minimal knowledge about the process involved. This might seem too much to ask for. However, as stated at the beginning of this section, the goal can be attained by using past experience to improve the performance. This is the key feature that distinguishes the ILC from conventional feedback controllers. From **P1-P6**, it can be observed that the ILC rationale is quite simple. The same process is operated repeatedly and each time the control signal is updated based on the feedback information of previous trials or repetitions. Eventually, the control signal is modified to a trajectory that when applied to the system it produces the desired output. Mathematically speaking, the control signal converges to the signal that corresponds to the inverse solution of the system for the desired output. Therefore, ILC can be viewed as an iterative inversion process. The direct utilization of real system data and past information in form of repetitions, allows the inversion to proceed without detail modeling.

It should be pointed out that the ILC is not an open-loop control operation, although the ILC only modifies the input command for the next repetition. ILC is closed-loop in the iteration number direction since updates are performed for the next repetition using the feedback measurements of the previous repetition. This is similar to the closed-loop structure of conventional controllers in time axis which updates the control signal of the next time step using the feedback at current or past time steps. The difference is that the thinking in ILC is in the repetition domain, which makes it appears as open-loop control in the time domain.

In practice, the ILC should be taken as an addition to the existing conventional controller. When analyzing the property of the ILC, considerations should be taken both in the time axis and in the repetition number direction, which is in essence in the category of the 2-D system theory. However, as the repetitive task is to be executed in a fixed finite time interval, more attentions should actually be paid in the repetition axis in the analysis of the iterative learning control property.

The following definitions on norms are used in this chapter and throughout this monograph:

$$\|f\| = \max_{1 \leq i \leq n} |f_i|,$$

$$\|G\| = \max_{1 \leq i \leq m} (\sum_{j=1}^{n} |g_{i,j}|),$$

$$\|h(t)\|_\lambda = \sup_{t \in [0,T]} e^{-\lambda t} \|h(t)\|,$$

where $f = [f_1, \cdots, f_n]^T$ is a vector, $G = [g_{i,j}]_{m \times n}$ is a matrix and $h(t)$ ($t \in [0,T]$) is a real function.

1.1.3 Literature Overview

It is interesting to note that around 1984, there are simultaneously five groups individually reporting the fact that the system repetition can be utilized to improve the system control performance. They are:

- S. Arimoto, S. Kawamura and F. Miyazaki [17],
- G. Casalino and G. Bartolini [36],
- J. J. Craig [70],
- R. H. Middleton, G. C. Goodwin and R. W. Longman [170],
- E. G. Harokopos, [99].

Among these, the term 'Iterative Learning Control' (ILC) coined by Arimoto and his associates is gradually widely accepted. The formalized theoretical issues were firstly investigated by Arimoto and his associates [15, 18].

Some overview descriptions of ILC research can be found in [175, 214, 153, 174, 172, 109, 186, 28]. A very recent expository overview is given by Moore [173]. A reference library for ILC research has been built by the first author with the co-operation of Prof. K. L. Moore. The library is a set of BIBTEX files containing most of the existing literatures on ILC for use in LATEX, which was announced in the E-letter # 110 (Oct. 1997) and is now publically accessible and downloadable through internet at

        http://ilc.ee.nus.edu.sg/ILC/ilcref.html

with a mirror at

    http://www.crosswinds.net/singapore/~yqchen/ILC/ilcref.html.

The ILC reference library is also a part of the *The Collection of Computer Science Bibliographies* located at

        http://liinwww.ira.uka.de/bibliography/Ai/ILC.html.

A list of hyperlinks to ILC research groups and individuals around the world

can also be found at
              http://ilc.ee.nus.edu.sg/ILC/ilclinks.html
mirrored at
    http://www.crosswinds.net/singapore/~yqchen/ILC/ilclinks.html.
There are more than 21 entries which are expected to grow in the future
as more and more attentions are being drawn to the research and applications of ILC and its peripherals. Inside many of the links, some of the recent
publications of respective researchers may be found on-line downloadable.

Instead of reviewing the literature as whole as the way in [173], in this
section, based on·the authors' observation, literatures related to some specific topics are briefly listed in the following. For detailed comments of the
listed topics, again, the authors recommend literature [173]. Detailed review
of some of the topics concerned in this monograph can be found in the introduction sections of respective chapters. Additional comments regarding
possible future development in ILC research can be found in Sec. 12.2.

- **D-type ILC and its robustness.**
  [17, 101, 102, 103, 49, 141, 228].
- **P-type ILC and its robustness.**
  [20, 159, 218, 227, 48, 215, 65, 78, 53].
- **H^∞ analysis and design.**
  [12, 72].
- **Frequency-domain analysis and design.**
  [156, 107, 81, 154, 82, 111, 104, 166, 123, 4].
- **Non-minimum phase problem.**
  [211, 6].
- **Initialization problem.**
  [143, 194, 195, 200, 103, 151, 139, 51, 59, 55, 64].
- **Tracking varying $y_d(t)$.**
  [214, 57, 58].
- **Constrained dynamical systems.**
  [19, 41, 42, 39, 177, 34, 40, 176, 244, 125, 229].
- **Time delays.**
  [49, 46, 210, 252, 106, 188].
- **Predictive ILC.**
  [258, 9, 57, 32, 146, 60].
- **Discrete-time and sampled data systems.**
  [235, 190, 87, 206, 114, 105, 118, 128, 219, 208, 136, 238, 42, 119, 82, 232, 216, 199, 217, 7, 13, 193, 240, 220, 57, 58, 76, 60, 77, 63].
- **Feedback controlled systems.**
  [100, 184, 159, 142, 185, 208, 145, 241, 144, 239, 120, 251, 209, 4, 10, 58, 67, 66, 96, 53, 63].
- **Linear 2D systems.**
  [30, 87, 86, 204, 90, 89, 136, 8, 216].

- **Nonlinear 2D theory.**
  [135].
- **Combined with existing control schemes.**
  [197, 198, 134, 129, 250].
- **High-order ILC.**
  [27, 49, 53, 46, 78, 76, 54, 56, 60].
- **Estimation/identification based ILC.**
  [181, 114, 142, 138, 166, 192, 147].
- **Adaptive ILC schemes.**
  [70, 71, 169, 221, 130, 168, 236, 111, 185, 198, 83, 122, 237, 226, 108].
- **ILC with a set of learned trajectories.**
  [22, 124, 125, 191, 248, 127, 253, 254].
- **Point-to-point ILC.**
  [158, 160, 162, 164, 163, 161, 165, 249].
- **Optimization based ILC.**
  [99, 256, 201, 11, 187].
- **Principle of self-support (PSS).**
  [179, 180].
- **Nonlinear ILC updating law.**
  [48, 50].
- **ILC applied to robotics.**
  See [173].
- **New domains of ILC application.**
  See [173].
- **Others.**
  See [173].

From the authors' observation, the future research topics in the area of ILC will be very active. Some of the considerations are described in Chapter 12. Some of the new aspects in ILC, including the promising fields of applications of ILC, will be presented in this monograph in the following chapters.

1.2 Objectives

The major objectives of ILC research are the stability or convergence and robustness analysis of learning algorithms. As most practical systems do not have a finite escape time, the analysis of ILC is mainly concentrated on the repetition direction. The stability of an ILC updating law (1.2) is to guarantee that for a certain function norm $\|e(\cdot)\|$

$$\|e_{k+1}\| \leq \|e_k\|, \text{for } k = 1, 2, \cdots \tag{1.3}$$

or, in a stronger sense,

$$\|e_{k+1}\| \leq \theta \|e_k\|, \text{for } k = 1, 2, \cdots \tag{1.4}$$

with a constant $\theta \in (0,1)$.

The investigation of the robustness of the ILC algorithm (1.2) is to study its convergence under the revised version of postulates **P2-P5** as given below:

- **P2'.** The desired output $y_d(t)$ can be allowed to change from cycle to cycle under the restriction that the variation is bounded.
- **P3'.** Repeatability of the initial setting is satisfied within an admissible deviation level, i.e., the initial state $x_k(0)$ of the objective system can be set as

$$x_k(0) = x^0 + \delta_k, \quad \|\delta_k\| \leq b_{x_0} \tag{1.5}$$

for some small constant $b_{x_0} \geq 0$.
- **P4'.** Invariance of the system dynamics may not be maintained and hence, the magnitude of the modelling error $w_k(t)$ satisfies

$$\sup_{t \in [0,T]} \|w_k(t)\| \leq b_w, \quad \forall k \tag{1.6}$$

for some small constant $b_w \geq 0$.
- **P5'.** For each iteration, the output $y_k(t)$ can be measured within a small specified noise level.

The ILC law (1.2) is said to be robust if under postulates **P2'-P5'**, the system stability along the iteration axis is still guaranteed. The stability here is in the sense that the final tracking error bound is a class-K function [117, page 109] of the assumed bounds in **P2'-P5'**. Furthermore, it is desirable to be able to adjust the final tracking error bound as well as the ILC convergence rate.

1.3 Preview of Chapters

There are 12 chapters in this monograph. Among the 12 chapters, two chapters (10 and 11) are dedicated to ILC applications and eight chapters (Chapters 2 - 9) to some new theoretical developments in ILC.

In Chapter 2, a high-order D-type ILC is proposed for an improved learning transient along the iteration number direction. A PID-type high-order iterative learning control algorithm is proposed for a class of delayed uncertain nonlinear systems. The convergence property of the proposed learning control algorithm is guaranteed when the re-initialization errors, the uncertainties and disturbances to the systems are bounded. It is shown that the final tracking error bound is a class-K function of the bounds of the reinitialization errors, the uncertainties and disturbances. This implies that the bounds of the errors between the desired and the actual trajectories of the systems will also tend to zero when the reinitialization errors, the uncertainties and disturbances decay to zero. It is also examined that the time delays in the system states do not play a significant role in the ILC convergence. The effectiveness of the high-order ILC is demonstrated by simulation examples.

In Chapter 3, a high-order P-type ILC scheme using the current iteration tracking error (CITE) is proposed for a class of continuous-time uncertain nonlinear repetitive systems. Uniform boundedness of the tracking error is established in the presence of uncertainty, disturbance and initialization error. Moreover, the tracking error bound and the ILC convergence rate are shown to be adjustable by tuning the learning gain of the CITE. Illustrative simulations are presented to validate the theoretical results.

In Chapters 4 and 5, discrete-time uncertain nonlinear systems are studied systematically. Chapter 4 presents an analysis of ILC using CITE for discrete-time uncertain nonlinear systems. A similar result to Chapter 3 has been obtained on the tuning of tracking error bound and ILC convergence rate. The result of Chapter 4 is helpful to understand the role of a feedback controller in iterative learning control. In Chapter 5, high-order ILC schemes with a general feedback controller are proposed and analysed. A feedback controller is shown to be helpful to improve the ILC convergence performance. The control input saturation is considered in the ILC analysis.

A commonly used assumption (**P3, P3'**) is that the initial states in each repetitive operation should be inside a given ball centered at the desired initial states which may be unknown. This assumption is critical to the stability analysis and the size of the ball will directly affect the final output trajectory tracking errors as shown in Chapter 2. In Chapter 6, it is shown that this assumption can be removed by using an iterative initial state learning scheme together with a traditional D-type ILC updating law. Both linear and nonlinear time-varying uncertain systems are investigated. Uniform bounds for the final tracking errors are obtained and these bounds are only dependent on the system uncertainties and disturbances, yet independent of the initial errors. Furthermore, the desired initial states can be identified through learning iterations. The results are extended to the high-order case. Simulation illustrations are given for both the first-order and the high-order schemes.

Unlike the conventional ILC, a special type of iterative learning control problem is considered in Chapter 7. Due to the insufficient measurement capability in many real control problems such as RTPCVD, it may happen that, instead of the whole output trajectory tracking error, only the terminal output tracking error is available. Focusing on a typical terminal thickness control problem in RTP chemical vapor deposition (CVD) of wafer fab industry, we propose a high-order Terminal Iterative Learning Control method. The ultimate control objective in RTPCVD is to control the deposition thickness (DT) at the end of an RTP cycle. The control profile for the next operation cycle has to be updated using the terminal DT tracking error alone, as the DT is measured only at the end of each RTP operation. By parameterizing the control profile with a piecewise continuous functional basis, the parameters are updated by a high-order updating scheme. A convergence condition is established for a class of uncertain discrete-time time-varying linear systems. Simulation results for an RTPCVD thickness control problem are presented to demonstrate the effectiveness of the proposed iterative learning scheme.

Chapters 2-7 concentrate on the *analysis* of ILC convergence and robustness issues for various learning algorithms. However, the ILC *design* issues have to be considered. Especially, to draw the attention from practitioners, an easy to use design method must be developed. In Chapter 8, a noncausal finite impulse respose (FIR) filtering based iterative learning controller is developed. Similarly, an ILC is applied as a feedforward controller to the existing feedback controller. The convergence analysis shows that it is alway possible to ensure the ILC convergence with a proper choose of two parameters: the learning gain and the filtering length. Therefore, the *design* task of ILC is reduced to tuning these two parameters. Detailed design procedures are explicitly given. Some practical issues on the parameter tuning are outlined. A limit on the ILC convergence rate has been obtained with some insightful discussions.

Chapter 9 presents a parallel results of Chapter 8 in the continuous-time domain using the idea of locally symmetrical double integral.

In Chapter 10, a new application of ILC to a robust curve identification problem - identifying the projectile's aerodynamic drag coefficient curve from tracking radar measured velocity data, is presented. Practical radar data given in a detailed table is used. A high-order ILC scheme is applied, similar to that of Chapter 2. The curve identification problem is equivalent to optimal tracking control problem with a minimax performance index. The successful solution presented in this chapter for such kind of problem is shown to be very effective. For a comparison, an alternative new method *Optimal Dynamic Fitting* developed by the authors is compared. It is shown that the ILC method is a much simpler yet more effective for curve identification applications.

In Chapter 11, another new application of ILC to the control of Functional Neuromuscular Stimulation Systems is presented. Simulation studies are performed with a practical musculoskeletal model given in detail. The discrete-time high-order ILC scheme discussed in Chapter 5 is applied. Extensive simulation studies show the effectiveness of the proposed ILC method and at the same time suggest that, compared to existing control methods, the ILC strategy may provide a better solution to the *customization, adaptation* and *robustness* problems in FNS system control. Moreover, it should be pointed out that unlike the conventional adaptive control method, the adaptation via ILC method is in a *point-wise* fashion. Thus, ILC is hopefully suitable for the real task execution both in the training stage use and in the daily-life use of the FNS systems.

Conclusions are drawn and some future research directions are highlighted in Chapter 12.

2. High-order Iterative Learning Control of Uncertain Nonlinear Systems with State Delays

2.1 Introduction

A COMMON OBSERVATION that human beings can learn perfect skills through repeated trials motivates the idea of iterative learning control (ILC) for systems performing repetitive tasks. Iterative learning control requires less *a priori* knowledge about the system dynamics and computational effort than many other kinds of control and has received a great deal of attention from many researchers in recent years [17, 70, 31, 27, 90, 225, 103, 3, 157, 120].

In iterative learning control, a fundamental problem is to guarantee the ILC convergence property, i.e., to guarantee the output trajectory of the system converging to the desired one within a prescribed accuracy as the number of ILC iterations increases. In early works such as [17] and [70], ILC was used for time-invariant mechanical systems and its convergence property was analyzed by using small signal method. The ILC convergent property was guaranteed under the assumption that the initial and all the subsequent trajectories are in a neighborhood of the desired trajectory. However, the existence and the size of the neighborhood were not investigated. By using global analysis method, the ILC convergence property with an arbitrary initial trajectory was shown in [101] and [31]. In [103], Heinzinger *et al.* studied the convergence of an iterative learning control algorithm for nonlinear time-varying systems with disturbances and uncertain reinitialization conditions. They showed that, under certain assumptions, the asymptotic tracking errors were bounded and the bounds were continuous functions of the bounds of the reinitialization errors as well as disturbances. By using the concept of relative degrees of dynamical systems, Ahn *et al.* [3] extended the result of [103] and proposed an ILC updating law which employed high order derivatives of tracking errors. Based on the existence of a direct transmission term in the system dynamics, the applicable system class which can be controlled by the ILC scheme has been well investigated [225]. Proposals have been made to combine ILC with existing feedback control schemes where feedback control stabilizes the system while ILC improves the system performance repeatedly [132, 157, 120], or with a parameter identification scheme to obtain an estimate of the learning parameters [181, 168, 110].

Most of the existing ILC schemes mentioned above are based on the first-order updating laws, i.e., only the information of one previous ILC iteration is employed. In [27], a high-order ILC algorithm is used for tracking control of time-varying linear systems, where, to construct the control in the current iteration, the information of several previous learning iterations, including control functions, tracking errors as well as their derivatives, is used. It was demonstrated in [27] that high-order ILC schemes did have potential to give a better convergence performance than the first-order ILC schemes. In fact, high-order ILC schemes can be used to improve the transient learning behavior along the learning iteration number direction. Consider the ILC updating law [17]

$$u_{i+1}(t) = u_i(t) + \Gamma \dot{e}_i(t)$$

along the ILC iteration number i direction where $e(t)$ is the tracking error, $u(t)$ is the control, and Γ is the learning gain matrix. Clearly, it is only an integral controller. By using the difference $\dot{e}_i(t) - \dot{e}_{i-1}(t)$ as the derivative approximation along i-direction, the PID controller in i-direction will result in the following form of the ILC updating law

$$u_{i+1}(t) = u_i(t) + \Gamma \dot{e}_i(t) + \Gamma_1 \dot{e}_{i-1}(t) + \Gamma_2 \dot{e}_{i-2}(t),$$

which is actually a high-order iterative learning controller. As we know that PID controllers can generally give better performance than integral controllers, thus we can expect that the high-order ILC is capable of giving better ILC performance than the traditional first-order ILC.

In this chapter, we consider a larger class of systems than that in [27] and present a high-order iterative learning control algorithm including PID information of the tracking errors for uncertain nonlinear systems with delays. It is shown that the convergence property of the proposed learning control algorithm is guaranteed when the reinitialization errors, the uncertainties and disturbances to the systems are bounded. In the case when the reinitialization errors, uncertainties and disturbances decay to zero, the bounds of the tracking errors also decay to zero, which extends the convergence results in the first-order ILC in [103] and [102]. The proposed high-order ILC algorithm is applied to a single link manipulator with direct joint torque control. From the simulation studies, better ILC convergence performance is observed. Moreover, it is shown that the uncertain time delays in state variables do not affect the ILC convergence property significantly.

This chapter is organized as follows. In Sec. 2.2, a class of nonlinear repetitive systems with delays in the state variables is introduced and the control problem is formulated with a high-order ILC updating law. A convergence analysis is performed in Sec. 2.3 where the ILC convergence condition is established. Simulation studies are presented in Sec. 2.4 to illustrate the effectiveness of the proposed ILC scheme together with a discussion on the ILC design issue. Finally, conclusions are drawn in Sec. 2.5.

2.2 High-order ILC Problem

Consider a time-varying delayed uncertain nonlinear system which performs a given task repeatedly as

$$\begin{cases} \dot{x}_i(t) = f(x_i(t), x_i(t - t_{d1}), t) + B(x_i(t), x_i(t - t_{d2}), t)u_i(t) + w(x_i(t), t) \\ y_i(t) = g(x_i(t), t) + v_i(t) \end{cases} \quad (2.1)$$

where i denotes the i-th repetitive operation of the system; $x_i(t) \in R^n$, $u_i(t) \in R^m$, and $y_i(t) \in R^r$ are the state, control input and output of the system, respectively; $w(\cdot, \cdot)$, $v_i(\cdot)$ are bounded uncertainties or disturbances to the system; $t \in [0, T]$ is the time and T is given; $t_{dj} \leq T \, (j = 1, 2)$ are unknown time delays. It is assumed that when $t < 0$, $x_i(t) = 0$. The functions $f : R^n \times R^n \times [0, T] \to R^n$ and $B : R^n \times R^n \times [0, T] \to R^{n \times m}$ are piecewise continuous in t, and $g : R^n \times [0, T] \to R^r$ is a C^2 function in $[0, T]$, i.e., the partial derivatives $g_x(\cdot, \cdot)$ and $g_t(\cdot, \cdot)$ are also differentiable in x and t.

Given a desired output trajectory $y_d(t)$, the control objective is to find a control input $u_i(t)$ such that when $i \to \infty$, the system output $y_i(t)$ will track the desired output trajectory $y_d(t)$ as close as possible. The control input $u_{i+1}(t)$ is obtained by

$$u_{i+1}(t) = U(I_i)$$

where U is an updating law and I_i represents the information of the previous iterations, which is available at the current $((i+1) - th)$ iteration and is given by

$$I_i = \{(u_j(t), y_j(t), y_d(t)) \, | \, (i - N + 1) \leq j \leq i\}$$

where the integer $N \, (N \geq 1)$ is the order of ILC algorithm.

Before presenting a high-order iterative learning control algorithm, we first impose the following assumptions on the system (2.1).

- (A1). The system (2.1) is causal. Specifically, for a given bounded desired output $y_d(t)$, there exists a unique bounded input $u_d(t)$, $t \in [0, T]$ such that when $u(t) = u_d(t)$, the system has a unique bounded state $x_d(t)$ and $y_d(t) = g(x_d(t), t), t \in [0, T]$.
- (A2). The functions f, B, w, g and the partial derivatives g_x, g_t are uniformly globally Lipschitzian in x on $[0, T]$, i.e., there exist constants $k_h, k_{f_j}, k_{B_j} (j = 1, 2)$ such that

$$\|h(x_1(t), t) - h(x_2(t), t)\| \leq k_h \, \|x_1(t) - x_2(t)\|,$$

$$\|f(x_1(t), x_1(t - t_{d1}), t) - f(x_2(t), x_2(t - t_{d1}), t)\|$$
$$\leq k_{f_1} \, \|x_1(t) - x_2(t)\| + k_{f_2} \, \|x_1(t - t_{d1}) - x_2(t - t_{d1})\|$$

$$\|B(x_1(t), x_1(t - t_{d2}), t) - B(x_2(t), x_2(t - t_{d2}), t)\|$$
$$\leq k_{B_1} \, \|x_1(t) - x_2(t)\| + k_{B_2} \, \|x_1(t - t_{d2}) - x_2(t - t_{d2})\|$$

where $h \in \{g, g_x, g_t, w\}$.

- (A3). The functions g_x, v_i, $\dot{v}_i$ and B are uniformly bounded. In the sequel, we use b_{g_x}, b_v, $b_{\dot{v}}$ and b_B to denote the upper bounds for g_x, v_i, $\dot{v}_i$ and B, respectively. For example,

$$b_B = \sup_{\forall (x_1, x_2) \in R^n \times R^n} \sup_{t \in [0,T]} \|B(x_1, x_2, t)\| .$$

Now, we propose the following high-order ILC updating law for the system (2.1) which uses the P, I, D information of tracking errors in a similar way to [102]:

$$u_{i+1}(t) = \sum_{k=1}^{N} (1 - \gamma) P_k(t) u_l(t) + \gamma u_0(t)$$

$$+ \sum_{k=1}^{N} \{ Q_k(y_l(t), t) e_l(t) + R_k(y_l(t), t) \dot{e}_l(t) + S_k(y_l(t), t) \int_0^t e_l(\tau) d\tau \} \ (2.2)$$

where $l = i - k + 1$, $e_l(t) = y_d(t) - y_l(t)$, integer $N \geq 1$ is the order of the ILC algorithm; Q_k, R_k and S_k are learning operators; γ $(0 \leq \gamma < 1)$ is a weighting parameter to restrain the large fluctuation of the control input at the beginning of ILC iterations. The learning operators Q_k, R_k, and S_k are chosen to be bounded and their upper bounds, denoted by b_Q, b_R and b_S respectively, are defined by, for example

$$b_Q = \max_{1 \leq k \leq N} \sup_{t \in [0,T]} \sup_{\forall y \in R^r} \|Q_k(y, t)\| .$$

The weighting parameter γ can be selected to be a monotonic decreasing function w.r.t. the ILC iteration times i. As usual, it is assumed that $e_j(t) = 0$ and $u_j(t) = 0$ for $j < 0$.

2.3 Convergence Analysis

The convergence condition for the above high-order ILC scheme is shown in the following theorem.

Theorem 2.3.1 *Consider the repetitive system (2.1) satisfying assumptions (A1)-(A3) and assume that the initial state bias $x_d(0) - x_i(0)$ is bounded. If*

$$\sum_{k=1}^{N} P_k(t) = I_m \tag{2.3}$$

and there exist positive numbers ρ_k satisfying

$$\|(1 - \gamma) P_k(t) - R_k g_x B\| \leq \rho_k, \quad \forall (x, t) \in R^n \times [0, T]$$

$$\sum_{k=1}^{N} \rho_k = \rho < 1 \tag{2.4}$$

then when $i \to \infty$, the bounds of the tracking errors $\| u_d(t) - u_i(t) \|$, $\| x_d(t) - x_i(t) \|$ and $\| y_d(t) - y_i(t) \|$ converge asymptotically to a residual ball centered at the origin. Additionally, when the bounds of the initial state bias, disturbances and uncertainties tend to zero, all the bounds of the tracking errors also tend to zero.

The following lemma is needed in the proof of Theorem 2.3.1.

Lemma 2.3.1 Suppose a real positive series $\{a_n\}_1^\infty$ satisfies

$$a_n \le \rho_1 a_{n-1} + \rho_2 a_{n-2} + \cdots + \rho_N a_{n-N} + \varepsilon, \quad (n = N+1, N+2, \cdots),$$

where $\rho_i \ge 0$, $(i = 1, 2, \cdots, N)$, $\varepsilon \ge 0$ and

$$\rho = \sum_{i=1}^N \rho_i < 1,$$

then the following holds:

$$\lim_{n \to \infty} a_n \le \varepsilon/(1 - \rho). \tag{2.5}$$

For the case $N = 1$, a proof can be found in [102]. An early version of proof of Lemma 2.3.1 can be found in [49]. A revised version is given as follows.

Proof. Let $n_1 \in \{n - 1, n - 2, \cdots, n - N\}$ be an index number such that

$$a_{n_1} = \max\{a_{n-1}, a_{n-2}, \cdots, a_{n-N}\}. \tag{2.6}$$

Then, by the assumptions in the Lemma 2.3.1,

$$\begin{aligned} a_n &\le \rho_1 a_{n-1} + \rho_2 a_{n-2} + \cdots + \rho_N a_{n-N} + \varepsilon \\ &\le \rho a_{n_1} + \varepsilon. \end{aligned} \tag{2.7}$$

Similarly, let $n_2 \in \{n_1 - 1, n_1 - 2, \cdots, n_1 - N\}$ such that

$$a_{n_2} = \max\{a_{n_1-1}, a_{n_1-2}, \cdots, a_{n_1-N}\},$$

then,

$$a_{n_1} \le \rho a_{n_2} + \varepsilon.$$

Therefore,

$$a_n \le \rho^2 a_{n_2} + \rho \varepsilon + \varepsilon. \tag{2.8}$$

In general, we have

$$\begin{aligned} a_n &\le \rho^m a_{n_m} + \rho^{m-1} \varepsilon + \rho^{m-2} \varepsilon + \cdots + \rho \varepsilon + \varepsilon. \\ &= \rho^m a_{n_m} + \frac{1 - \rho^m}{1 - \rho} \varepsilon, \end{aligned}$$

where m and n_m are positive integers. If m is chosen such that $n_m \leq N$, then $[n/N] - 1 \leq m \leq n - N$, and therefore $m \to \infty$ when $n \to \infty$. Let $M = \max\{a_1, a_2, \cdots, a_N\}$, then

$$a_n \leq \rho^m M + \frac{1 - \rho^m}{1 - \rho} \varepsilon, \tag{2.9}$$

which implies

$$\lim_{n \to \infty} a_n \leq \frac{\varepsilon}{1 - \rho}. \tag{2.10}$$

This completes the proof of Lemma 2.3.1.

For the sake of brevity, the following notations will be used:

$$h_i = h(x_i(t), t); \quad h_d = h(x_d(t), t);$$

$$\delta h_i = h_d - h_i; \quad b_h = \sup_{t \in [0,T], x \in R^n} \|h(x, t)\|.$$

where h represents a function concerned. The partial derivatives of $h(x, t)$ are denoted by

$$h_{xi} = \frac{\partial h(x, t)}{\partial x} \bigg|_{x = x_i(t)}, \quad h_{ti} = \frac{\partial h(x, t)}{\partial t} \bigg|_{x = x_i(t)},$$

$$h_{xd} = \frac{\partial h(x, t)}{\partial x} \bigg|_{x = x_d(t)}, \quad h_{td} = \frac{\partial h(x, t)}{\partial t} \bigg|_{x = x_d(t)}.$$

Let k_h be the Lipschitz constant of the function h w.r.t x in $[0, T]$, then it is easy to see that

$$\|\delta h_i\| \leq k_h \|\delta x_i\|.$$

We now proceed to present a proof of Theorem 2.3.1.

Proof. From (2.1), it is easy to see that

$$e_l \overset{\triangle}{=} y_d - y_l = \delta g_l - v_l$$

$$\dot{e}_l = \dot{y}_d - \dot{y}_l = g_{xd}\dot{x}_d + g_{td} - g_{xl}\dot{x}_l - g_{tl} - \dot{v}_l$$

$$= \delta g_{xl}\dot{x}_d + g_{xl}\delta\dot{x}_l + \delta g_{tl} - \dot{v}_l$$

where $l = i, i - 1, \cdots, i - N + 1$, and

$$\dot{x}_d = f_d + B_d u_d + w_d$$

$$\delta\dot{x}_l = f_d + B_d u_d + w_d - f_l - B_l u_l - w_l$$

$$= \delta f_l + \delta B_l u_d + B_l \delta u_l + \delta w_l$$

and $\delta f_l = f_d - f_l = f(x_d(t), x_d(t - t_{d1}), t) - f(x_l(t), x_l(t - t_{d1}), t)$, $\delta B_l = B_d - B_l = B(x_d(t), x_d(t - t_{d2}), t) - B(x_l(t), x_l(t - t_{d2}), t)$. Now, by using ILC updating law (2.2) and the condition (2.3), one gets

$$\delta u_{i+1} = u_d - u_{i+1} = \sum_{k=1}^{N}(1-\gamma)P_k(t)\delta u_l + \gamma\delta u_0$$

$$-\sum_{k=1}^{N}Q_k(y_l,t)(\delta g_l - v_l) - \sum_{k=1}^{N}S_k(y_l,t)\int_0^t(\delta g_l - v_l)\mathrm{d}\tau$$

$$-\sum_{k=1}^{N}R_k(y_l,t)\{\delta g_{xl}\dot{x}_d + \delta g_{tl} - \dot{v}_l + g_{xl}(\delta f_l + \delta w_l + B_l\delta u_l + \delta B_l\,u_d)\}$$

$$=\sum_{k=1}^{N}[(1-\gamma)P_k(t) - R_k(y_l,t)g_{xl}B_l]\delta u_l + \gamma\delta u_0$$

$$-\sum_{k=1}^{N}Q_k(y_l,t)(\delta g_l - v_l) - \sum_{k=1}^{N}S_k(y_l,t)\int_0^t(\delta g_l - v_l)\mathrm{d}\tau$$

$$-\sum_{k=1}^{N}R_k(y_l,t)\{\delta g_{xl}\dot{x}_d + g_{xl}(\delta f_l + \delta w_l + \delta B_l\,u_d) + \delta g_{tl} - \dot{v}_l\}. \quad (2.11)$$

Taking norms yields

$$\|\delta u_{i+1}\| \le \sum_{k=1}^{N}\rho_k\,\|\delta u_l\| + \gamma\,\|\delta u_0\| + \sum_{k=1}^{N}b_Q(k_g\,\|\delta x_l\| + b_v)$$

$$+\sum_{k=1}^{N}b_S\int_0^t(k_g\,\|\delta x_l\| + b_v)\mathrm{d}\tau + \sum_{k=1}^{N}b_R\{k_{g_x}b_{\dot{x}_d}\,\|\delta x_l\|$$

$$+k_{g_t}\,\|\delta x_l\| + b_{g_x}(k_{f_1} + k_w + k_{B_1}b_{u_d})\,\|\delta x_l\| + b_{\dot{v}}$$

$$+b_{g_x}k_{f_2}\|\delta x_l(t-t_{d_1})\| + b_{g_x}k_{B_2}b_{u_d}\|\delta x_l(t-t_{d_2})\|\}$$

$$\le \sum_{k=1}^{N}\rho_k\,\|\delta u_l\| + c_0 + \gamma\,\|\delta u_0\| + a_0\sum_{k=1}^{N}\|\delta x_l\| + a_1\sum_{k=1}^{N}\int_0^t\|\delta x_l\|\,\mathrm{d}\tau$$

$$+b_R\sum_{k=1}^{N}\{b_{g_x}k_{f_2}\|\delta x_l(t-t_{d_1})\| + b_{g_x}k_{B_2}b_{u_d}\|\delta x_l(t-t_{d_2})\|\}, \quad (2.12)$$

where $b_{u_d} = \sup_{t\in[0,T]}\|u_d(t)\|$, and

$a_0 = b_Q k_g + b_R[k_{g_x}b_{\dot{x}_d} + k_{g_t} + b_{g_x}(k_{f_1} + k_w + k_{B_1}b_{u_d})]$,

$a_1 = b_S k_g$,

$c_0 = N(b_Q b_v + b_S b_v T + b_R b_{\dot{v}})$.

From (2.1), the integral expression of $x_l(t)$

$$\|\delta x_l\| = \|\delta x_l(0) + \int_0^t \delta\dot{x}_l\,\mathrm{d}\tau\|$$

$$\le \|\delta x_l(0)\| + \int_0^t[(k_{f_1} + k_{B_1}b_{u_d} + k_w)\,\|\delta x_l\|$$

$$+k_{f_2}\|\delta x_l(t-t_{d_1})\| + k_{B_2}b_{u_d}\|\delta x_l(t-t_{d_2})\| + b_B\,\|\delta u_l\|]\mathrm{d}\tau. \quad (2.13)$$

Starting from (2.13), a relationship between $\|x_l(t)\|_\lambda$ and $\|u_l(t)\|_\lambda$ is to be found for the ILC convergence proof. First, for any function $x(t) \in R^n$, $t \in [0,T]$, the λ-norm for $\int_0^t \|x(\tau)\| \mathrm{d}\tau$ is

$$\sup_{t\in[0,T]} e^{-\lambda t} \int_0^t \|x(\tau)\| \mathrm{d}\tau = \sup_{t\in[0,T]} e^{-\lambda t} \int_0^t \|x(\tau)\| e^{-\lambda\tau} e^{\lambda\tau} \mathrm{d}\tau$$

$$\leq \|x(t)\|_\lambda \sup_{t\in[0,T]} e^{-\lambda t} \int_0^t e^{\lambda\tau} \mathrm{d}\tau$$

$$= \|x(t)\|_\lambda \sup_{t\in[0,T]} \frac{1 - e^{-\lambda t}}{\lambda}$$

$$\leq \|x(t)\|_\lambda O(\lambda^{-1}) \tag{2.14}$$

where

$$O(\lambda^{-1}) \triangleq \frac{1 - e^{-\lambda T}}{\lambda}. \tag{2.15}$$

Due to the fact that $\|x(t - t_d)\|_\lambda \leq \|x(t)\|_\lambda$, by referring to (2.14), one can find that

$$\sup_{t\in[0,T]} e^{-\lambda t} \int_0^t \|x(\tau - t_d)\| \mathrm{d}\tau \leq \|x(t)\|_\lambda O(\lambda^{-1}) e^{-\lambda t_d}. \tag{2.16}$$

By using the relationship (2.14), (2.16), (2.13) becomes

$$\|\delta x_l(t)\|_\lambda \leq \|x_l(0)\| + a_2 O(\lambda^{-1})\|\delta x_l(t)\|_\lambda + b_B O(\lambda^{-1})\|\delta u_l(t)\|_\lambda \tag{2.17}$$

where

$$a_2 \triangleq k_{f_1} + k_{B_1} b_{u_d} + k_w + k_{f_2} e^{-\lambda t_{d1}} + k_{B_2} b_{u_d} e^{-\lambda t_{d2}}.$$

Clearly, $\exists \lambda$, such that

$$a_2 O(\lambda^{-1}) < 1. \tag{2.18}$$

Thus

$$\|\delta x_l(t)\|_\lambda \leq \frac{\|x_l(0)\| + b_B O(\lambda^{-1})\|\delta u_l(t)\|_\lambda}{1 - a_2 O(\lambda^{-1})}. \tag{2.19}$$

Now performing the λ-norm operation for (2.12), by using (2.14), one obtains

$$\|\delta u_{i+1}(t)\|_\lambda \leq \sum_{k=1}^N \rho_k \|\delta u_l(t)\|_\lambda + c_0 + \gamma\|\delta u_0\|_\lambda + a_3 \sum_{k=1}^N \|\delta x_l(t)\|_\lambda \tag{2.20}$$

where

$$a_3 \triangleq a_0 + a_1 O(\lambda^{-1}) + b_R b_{g_*} (k_{f_2} e^{-\lambda t_{d1}} + k_{B_2} b_{u_d} e^{-\lambda t_{d2}}).$$

Substituting (2.19) to (2.20) yields

$$\|\delta u_{i+1}\|_\lambda \leq \sum_{k=1}^{N} \bar{\rho}_k \, \|\delta u_l\|_\lambda + \varepsilon \qquad (2.21)$$

where

$$\bar{\rho}_k \triangleq \rho_k + a_3 \frac{b_B O(\lambda^{-1})}{1 - a_2 O(\lambda^{-1})}, \qquad (2.22)$$

$$\varepsilon \triangleq c_0 + \gamma\|\delta u_0\|_\lambda + a_3 \sum_{k=1}^{N} \frac{\|x_l(0)\|}{1 - a_2 O(\lambda^{-1})}. \qquad (2.23)$$

By the condition (2.4) in Theorem 2.3.1, one can find a sufficiently large λ such that $\bar{\rho}_k < 1$ and $\bar{\rho} = \sum_{k=1}^{N} \bar{\rho}_k < 1$. Then, according to Lemma 2.3.1, it can be concluded that

$$\lim_{i\to\infty} \|\delta u_i(t)\|_\lambda \leq \frac{1}{1-\bar{\rho}}\varepsilon. \qquad (2.24)$$

From (2.19), (2.23) and that

$$\|\delta y_i\|_\lambda \leq k_g \, \|\delta x_i\|_\lambda + b_v, \qquad (2.25)$$

one can observe that the tracking errors $\|\delta u_i(t)\|$, $\|\delta x_i(t)\|$, and $\|\delta y_i(t)\|$ are all proportional to the initial state error $\|\delta x_i(0)\|$ and output disturbance bound by the definition of ε in (2.23). Moreover, if γ, $\|\delta x_i(0)\|$ and b_v tend to zero, through ILC repetitive operations, all the tracking error bounds tend to zero asymptotically, too. This completes the proof of Theorem 2.3.1.

Remark 2.3.1. Under the conditions (2.3) and (2.4), one knows that the ILC convergence is not affected by the uncertainties and disturbances, the initial state bias, neither by the selections of γ. But the bounds of the ILC final tracking errors are directly affected by those factors. As a high-order version ILC updating law of *[103]*, it utilizes the past experiences comprehensively and has more flexibilities in choosing learning operators and parameters. Hence, the better ILC performance can be expected. If the system dynamics is totally unknown, like the selection of learning parameters in traditional ILC algorithms *[17, 103]*, the order N selection is also based on a *trial-and-error* method, which should be also in an 'iterative learning' way. In practice, N is normally less than or equal to 3.

Remark 2.3.2. Iterative learning control method itself can not reject or compensate the uncertainties or disturbances or initial state bias. But, once they will not appear any more in the coming ILC iterations, the ILC can recover from their influence and achieve a perfect tracking.

Remark 2.3.3. The proportional and integral (PI) components in the ILC updating law do not affect the ILC convergence as also discussed in [102]. As demonstrated in [139], PI components will also be directly related to the robustness performance of ILC w.r.t. bounded initial positioning errors, uncertainties and disturbances while the derivative component governs the ILC convergence property. Conceptually, the PI learning operators can be used as design factors to make better ILC performance when there exist bounded initial errors and uncertainties. High-order ILC updating law can be regarded as the PID controller in the i-direction where the higher order information is used to approximate the D information in the i-direction. So, the *dual* PIDs both in the time t-direction and in the i-direction can make the ILC application easier by taking the advantage of the long-history PID usage in conventional control engineering.

2.4 Simulation Studies

2.4.1 Example 1: Effects of Time-delays in State-variables

A pure mathematical SISO system is used for the illustrative purpose only as follows:

$$\begin{cases} \dot{x}(t) = -2[1 + 0.2x(t) + 0.8x(t - \tau_1)]^{(0.2x(t)+0.8x(t-\tau_2))}\tanh(u(t)) \\ y(t) = x(t) \end{cases}$$

where τ_1, τ_2 are time-delays; $t \in [0, 1]$. The desired trajectory is

$$y_d(t) = \begin{cases} t, & t \in [0, 0.2]; \\ -t + 0.4, & t \in (0.2, 0.4]; \\ 0, & t \in (0.4, 0.6]; \\ t - 0.6, & t \in (0.6, 0.8]; \\ -t + 1, & t \in (0.8, 1.0]. \end{cases}$$

The simplest first-order D-type ILC updating law is used with a constant learning gain -0.2. In the simulation, time step is chosen as 0.01; $x_i(t)|_{t \leq 0} = 0$, $u_i(0) = 0, \forall i$. No disturbance or model uncertainty is assumed. The ILC stops when $e_b \triangleq \sup_{t \in [0,1]} |y_d(t) - y_i(t)| \leq 0.005$. With several different choices of τ_1 and τ_2, the obtained simulation results are summarized in Table 2.1.

It can be clearly observed that the time delays in state variables do not affect the ILC convergence significantly as analyzed in Theorem 2.3.1. Moreover, it is interesting to note that the nonlinear mathematical system considered in this simulation example is beyond the system class discussed in Theorem 2.3.1. This implies that ILC method is more generally applicable.

Table 2.1. Comparison of ILC convergence histories ($e_b - i$) with different state delays

ILC $\# i$	$\tau_1 = 0$ $\tau_2 = 0$	$\tau_1 = 0.3$ $\tau_2 = 0.6$	$\tau_1 = 0.6$ $\tau_2 = 0.3$	$\tau_1 = 0.2$ $\tau_2 = 0.8$	$\tau_1 = 0.8$ $\tau_2 = 0.2$
1	2.00000e-1	2.00000e-1	2.00000e-1	2.00000e-1	2.00000e-1
2	1.21632e-1	1.21835e-1	1.21836e-1	1.21775e-1	1.21776e-1
3	7.64010e-2	7.71088e-2	7.71150e-2	7.69006e-2	7.69053e-2
4	4.96064e-2	5.06872e-2	5.06989e-2	5.03696e-2	5.03783e-2
5	3.30745e-2	3.43140e-2	3.43290e-2	3.39503e-2	3.39616e-2
6	2.25227e-2	2.37635e-2	2.37795e-2	2.34005e-2	2.34125e-2
7	1.56154e-2	1.67662e-2	1.67817e-2	1.64310e-2	1.64425e-2
8	1.10091e-2	1.20275e-2	1.20414e-2	1.17326e-2	1.17430e-2
9	7.89450e-3	8.76818e-3	8.78027e-3	8.51747e-3	8.52637e-3
10	5.76562e-3	6.49984e-3	6.50999e-3	6.29162e-3	6.29897e-3
11	4.29698e-3	4.90592e-3	4.91429e-3	4.73596e-3	4.74186e-3

2.4.2 Example 2: Effects of High-order ILC schemes.

To demonstrate the effectiveness of the proposed high-order ILC algorithm for the improvement of the ILC convergence property, a single link direct joint driven manipulator model is used for the simulation study. The dynamics of the system is described by

$$\ddot{\theta}(t) = \frac{1}{J}(u(t) - F(t)) + \frac{1}{J}(\frac{1}{2}m + M)gl \sin \theta(t) \tag{2.26}$$

where $\theta(t)$ is the angular position of the manipulator; $u(t)$ is the applied joint torque; $F(t)$ is the friction torque; m, l are the mass and length of the manipulator respectively, M is the mass of the tip load, g is the gravitational acceleration and the J is the moment of inertia w.r.t the joint and is given by $J = Ml^2 + ml^2/3$. The parameters used in the simulation studies are listed in Table 2.2.

Table 2.2. Physical Parameters of A Single-link Manipulator

parameter	m	M	l	g
unit	kg	kg	m	m/sec.^{2}
value	2	4	0.5	9.8

The desired tracking trajectories are specified as

$$\theta_d(t) = \theta_b + (\theta_b - \theta_f)(15\tau^4 - 6\tau^5 - 10\tau^3) \tag{2.27}$$

$$\dot{\theta}_d(t) = (\theta_b - \theta_f)(60\tau^3 - 30\tau^4 - 30\tau^2) \tag{2.28}$$

where $\tau = t/(t_f - t_0)$. In the simulation, $\theta_b = 0°$, $\theta_f = 90°$, $t_0 = 0$, $t_f = 1$. The RK-4 method is used to numerically integrate the state equation with constant time step $h = 0.01$ second. The initial states at each ILC repetition are all set to 0. The ILC ends when $e_{b1} \leq 1°$ and at the same time when $e_{b2} \leq$

$1°/s$ where $e_{b1} \triangleq \sup_{t \in [0,1]} |\theta_d(t) - \theta(t)|$ and $e_{b2} \triangleq \sup_{t \in [0,1]} |\dot{\theta}_d(t) - \dot{\theta}(t)|$. The following high-order ILC updating law is used

$$u_{i+1}(t) = u_i(t) + \sum_{k=1}^{N} R_k \dot{e}_{i-k+1}(t) \tag{2.29}$$

where R_k is the learning parameter and $e(t) = \dot{\theta}_d(t) - \dot{\theta}(t)$. By using Theorem 2.3.1, it can be verified that when $N = 1$ the best choice of R_1 is J which makes $\rho = 0$. Assume that that the accurate system dynamics is not known and the estimated $\hat{J}$ is a quarter of the true value in the simulations.

To apply the high-order ILC scheme, the following three cases are considered

- Case 1: $N = 1$, $R_1 = \hat{J} = J/4$;
- Case 2: $N = 2$, $R_1 = J/4$, $R_2 = J/16$;
- Case 3: $N = 3$, $R_1 = J/4$, $R_2 = J/16$, $R_3 = J/16$.

The guideline of choosing R_1, R_2, and R_3 is given by (2.4) in Theorem 2.3.1. To guarantee the ILC convergence, ρ should be less than one. The above choice of R_1, R_2, and R_3 in Cases 1 to 3 clearly satisfies the convergence condition. Two situations related to the friction torque are considered. The simulation results of the above three cases, summarized in Figs. 2.1 and 2.2, are presented to illustrate the effectiveness of high-order ILC scheme over the conventional first-order one.

- Situation 1. No friction torque is considered, i.e., $F(t) = 0$.
  The solid lines are for the maximal absolute tracking error of angular velocities while the dotted lines for angular positions. For the same given accuracy, it can be noted that the total numbers of ILC repetitions required for the cases that $N = 1, 2$, and 3 are $26, 19$, and 15 respectively. An improved ILC performance can be clearly observed with a higher order ILC updating law.
- Situation 2. The friction torque $F(t)$ is that

$$F(t) = \begin{cases} f^+ + B^+ \dot{\theta}, & \dot{\theta} \geq 0, \\ f^- + B^- \dot{\theta}, & \dot{\theta} < 0, \end{cases}$$

where the Coulomb frictions $f^+ = 8.43$Nm and $f^- = -8.26$Nm; the viscous friction coefficients $B^+ = 4.94$Nm/rad/s and $B^- = 3.45$Nm/rad/s [68, page 257]. All the 3 cases as in the first situation are examined.
Clearly, the obtained results are similar to those in situation 1 and are presented in Fig. 2.2. As the viscous friction has some dissipative effect on the nonlinear system, the responses in this situation are more damped than those in Situation 1. Again, improved ILC performance can be observed as a higher ILC scheme is applied.

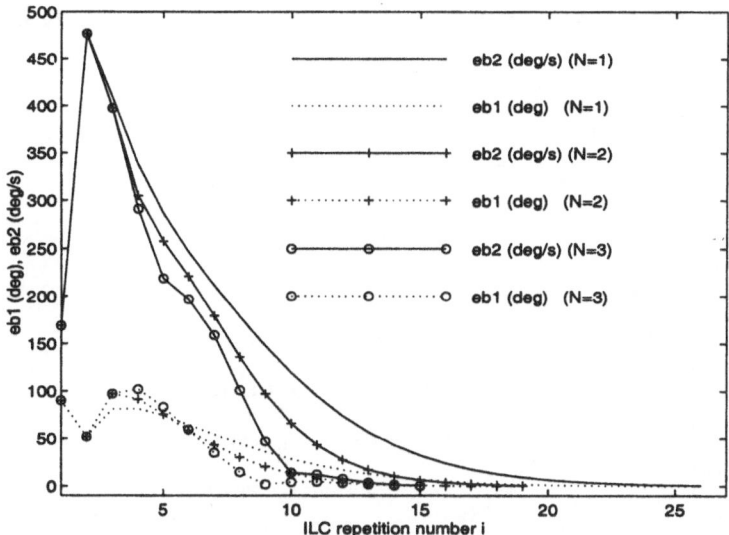

Fig. 2.1. Tracking performances comparison for $N = 1, 2, 3$, with $F(t) = 0$.

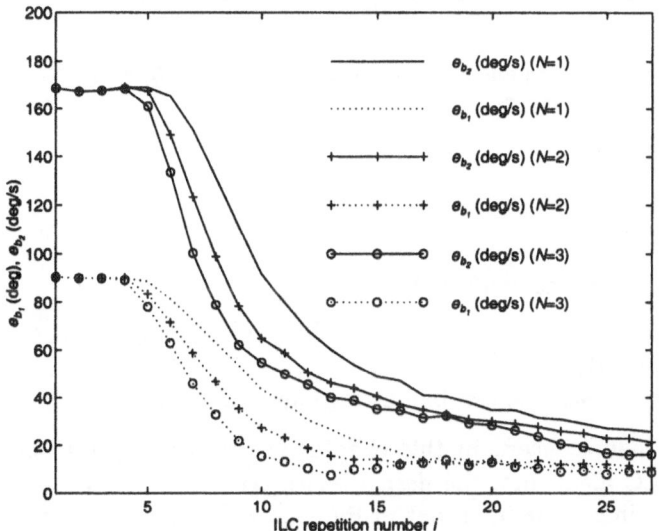

Fig. 2.2. Tracking performances comparison for $N = 1, 2, 3$, with $F(t) \neq 0$.

- Situation 3. $F(t) = 0$, with random re-initialization errors.
  Similar to Situation 2, in this situation, random re-initialization errors are added at the beginning of each iteration. The random re-initialization errors are set to (rand-0.5)*0.5 (m. and m/sec.), where rand is a MATLAB function to generate a uniformly distributed random noise over [0,1]. According to Theorem 2.3.1, the final tracking errors in this situation are

governed by the bounds of initialization errors. This is clearly shown in Fig. 2.3.

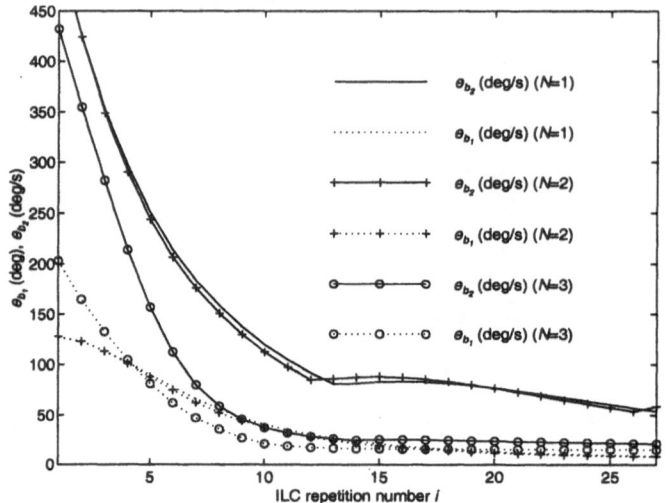

Fig. 2.3. Tracking performances comparison for $N = 1, 2, 3$, with $F(t) = 0$ and random re-initialization errors.

2.4.3 Discussions

The above two examples verified that

- time delays in state variables do not affect the ILC convergence performance significantly;
- a high-order ILC scheme can be better than a first-order one.

While the main concern in this chapter is the convergence *analysis* of a high-order ILC algorithm, the *design* issues are still open. Theorem 2.3.1 is only a guide line. As pointed out in Remark 2.3.1, the design task highly depends on the amount of knowledge of the system as well as the performance index imposed. If $N \leq 3$, from PID control point of view, the *design* can be made easy if one applies the PID setting and tuning ideas in conventional industrial control, as indicated by Remark 2.3.3. Chapter 8 emphasizes this ILC *design* issue and an interesting application with detailed design steps is given in [234].

2.5 Conclusion

The convergence of a PID-type high-order ILC algorithm for uncertain non-linear systems with state delays is shown to be guaranteed if the reinitialization errors, the uncertainties or disturbances are bounded. From a new point of view, i.e., the consideration in the iteration number direction, the ILC performance improvement by using high-order ILC algorithm is discussed and illustrated. The delays in the system states do not play significant roles in the ILC convergence analysis which is illustrate by a simulation example.

3. High-order P-type Iterative Learning Controller Using Current Iteration Tracking Error

3.1 Introduction

THE ILC UPDATING LAW discussed in Chapter 2 utilizes the derivative information of the tracking error. This is the so-called 'D-type' ILC. When the derivative and integral of the tracking error are not employed, the ILC updating law is referred to as the 'P-type'. The P-type scheme is more attractive for its simple structure, easy implementation and reduced sensory cost. Based on passivity properties of nonlinear systems, the robustness of a P-type ILC algorithm was shown in [16, 215]. It can be noted in the existing results that the boundedness of the final tracking error can be guaranteed in the presence of uncertainty, disturbance, and initialization error.

Nevertheless, it is not enough to ensure only the boundedness of the final tracking error. How to adjust the final tracking error bound is an important issue and has not been well studied yet. In this chapter, this problem is addressed by including the current iteration tracking error (CITE) in the ILC updating law. The concept of CITE in the ILC updating law was introduced in [257, 132, 184] where uncertainty, disturbance, and initialization errors were not considered. A high-order P-type ILC updating law utilizing CITE is proposed for a more general class of continuous-time uncertain nonlinear systems in this chapter. Uniform boundedness of the tracking error is established in the presence of uncertainty, disturbance, and re-initialization errors. Furthermore, it is shown that the final tracking error bound and the ILC convergence rate can be adjusted to a prescribed level by tuning the learning gain of the CITE.

The remaining parts of this chapter are organized as follows. The system description, the control objective and the proposed ILC updating law with CITE are presented in Sec. 3.2. Some notations are also given. An analysis of ILC convergence is performed in Sec. 3.3. It is shown from the analysis that the CITE gain can be used to tune the ILC convergence rate and the tracking error bound.

This is also illustrated by some simulation results given in Sec. 3.4. In Sec. 3.5, conclusion is given.

3.2 System Description and ILC Updating Law

Consider a repetitive nonlinear time-varying system with uncertainty and disturbance as follows.

$$\begin{cases} \dot{x}_i(t) = f(x_i(t), u_i(t), t) + w_i(t) \\ y_i(t) = g(x_i(t), t) + D(t)u_i(t) + v_i(t) \end{cases} \tag{3.1}$$

where i denotes the i-th iteration of the system's repetitive operation; $x_i(t) \in R^n$, $u_i(t) \in R^m$, and $y_i(t) \in R^r$ are the state, control input, and output of the system, respectively; the functions $f(\cdot, \cdot, \cdot) : R^n \times R^m \times [0, T] \mapsto R^n$ and $g(\cdot, \cdot) : R^n \times [0, T] \mapsto R^r$ are piecewise continuous and satisfy the Lipschitz continuity conditions, i.e., $\forall t \in [0, T]$,

$$\|f(x_{i+1}(t), u_{i+1}(t), t) - f(x_i(t), u_i(t), t)\| \le k_f (\|x_{i+1}(t) - x_i(t)\|$$
$$+ \|u_{i+1}(t) - u_i(t)\|),$$
$$\|g(x_{i+1}(t), t) - g(x_i(t), t)\| \le k_g (\|x_{i+1}(t) - x_i(t)\|) \tag{3.2}$$

where k_f, $k_g > 0$ are the Lipschitz constants; $w_i(t)$, $v_i(t)$ are uncertainty or disturbance to the system; $t \in [0, T]$ is the time and T is given; $D(t)$ is a time-varying matrix with appropriate dimension.

To restrict our discussion, the following assumptions are imposed on the class of systems described by (3.1).

- A1). The differences of the uncertainties, disturbances, and initialization errors between two successive repetitions are bounded as follows.

$$\|w_{i+1}(t) - w_i(t)\| \le b_w, \quad \|v_{i+1}(t) - v_i(t)\| \le b_v, \quad t \in [0, T], \; \forall i,$$
$$\|x_{i+1}(0) - x_i(0)\| \le b_{x0}, \quad \forall i,$$

  where b_{x0}, b_w, and b_v are unknown positive constants.
- A2). There exists some matrix $Q_0(t) \in R^{r \times m}$ such that $| I_r + D(t)Q_0(t) | \ne 0, \quad \forall t \in [0, T]$.
- A3). $D(t) \ne 0, \; t \in [0, T]$.

Remark 3.2.1. Assumption A1) only requires the boundedness of the differences between the uncertainties, disturbances, and initialization errors in two successive repetitions. Clearly, A2) can also be easily satisfied. Regarding A3), if $D(t) = 0$, the form (3.1) of the system description can still be obtained by differentiating the output equation, as discussed in [225].

Remark 3.2.2. Based on passivity properties of nonlinear system dynamics, the P-type ILC convergence property for affine nonlinear uncertain systems

$$\begin{cases} \dot{x}_i(t) = f(x_i(t), t) + B(x_i(t), t)u_i(t) + w_i(t) \\ y_k(t) = Cx_i(t) + v_i(t) \end{cases} \tag{3.3}$$

where C is a constant matrix satisfying some restrictions, has been established in [215]. We believe that it is possible to remove A3) by using the techniques in [215].

Remark 3.2.3. The class of system (3.1) under consideration is larger than that studied in [27] and [225] where the state equations were in an affine form.

Now the problem is formulated as follows. For a given realizable desired output trajectory $y_d(t)$, starting from an arbitrary continuous initial control input $u_0(t)$ and an arbitrary initial state $x_0(0)$, propose an ILC updating law to generate the next control input and the subsequent input series $\{u_i(t) \mid i = 2, 3, \cdots\}$ iteratively in such a way that when $i \to \infty$, $y_i(t) \to y_d(t) \pm \varepsilon^*$ and the bound ε^* is adjustable by tuning the learning gains of the ILC updating law.

To solve this problem, the following P-type ILC updating law utilizing the current iteration tracking error is employed.

$$u_{i+1}(t) = u_i(t) + \sum_{k=0}^{N} Q_k(t)e_l(t) \qquad (3.4)$$

where $l \triangleq i - k + 1$; $e_l(t) \triangleq y_d(t) - y_l(t)$ denoting the tracking error at l-th repetition; N is a positive integer denoting the order of the ILC updating law; and $Q_k(t), k = 0, 1, 2, \cdots, N$, are the learning gain matrices. Clearly, the current iteration tracking error $e_{i+1}(t)$ is used in the ILC updating law.

The convergence analysis of the ILC system (3.1) and (3.4) is given in the next section. For a brief presentation, the following notations will be used.

$$\Delta h_i(t) \triangleq h_i(t) - h_{i-1}(t), \qquad h_i(t) \in \{x_i(t), u_i(t), w_i(t), v_i(t)\},$$
$$\Delta f_i \triangleq f(x_i(t), u_i(t), t) - f(x_{i-1}(t), u_{i-1}(t), t),$$
$$\Delta g_i \triangleq g(x_i(t), t) - g(x_{i-1}(t), t),$$
$$b_D \triangleq \sup_{t \in [0,T]} \|D(t)\|,$$
$$b_{Q_k} \triangleq \sup_{t \in [0,T]} \|Q_k(t)\|, \quad k = 0, 1, \cdots, N,$$
$$O(\lambda^{-1}) \triangleq (1 - e^{-\lambda T})/\lambda \le 1/\lambda,$$
$$O_1(\lambda^{-1}) \triangleq k_f k_g O(\lambda^{-1})/[1 - k_f O(\lambda^{-1})],$$
$$\alpha_1 \triangleq \sup_{t \in [0,T]} \|I_r - D(t)Q_1(t)\|,$$
$$\alpha_k \triangleq \sup_{t \in [0,T]} \|D(t)Q_k(t)\|, \quad k = 2, 3, \cdots, N,$$
$$\eta \triangleq \sup_{t \in [0,T]} \|(I_r + D(t)Q_0(t))^{-1}\|,$$
$$\beta \triangleq \eta/(1 - \eta O_1(\lambda^{-1})b_{Q_0}).$$

3.3 Convergence Analysis

The robust convergence of the proposed high-order P-type iterative learning control algorithm utilizing the current iteration tracking error is presented in the following theorem.

Theorem 3.3.1 *Consider the ILC updating law (3.4) applied to the repetitive nonlinear uncertain system (3.1) satisfying Assumptions A1)-A3). For a given realizable desired trajectory $y_d(t)$ on the fixed time interval $[0, T]$, under the condition that*

$$\beta \sum_{k=1}^{N} \alpha_k = \rho < 1, \tag{3.5}$$

there exists a sufficiently large positive constant λ such that

- *1). the final tracking error is bounded and the bound is given by*

$$\lim_{i \to \infty} \|e_i(t)\|_\lambda \leq \frac{\varepsilon}{1 - \rho - \beta \sum_{k=1}^{N} O_1(\lambda^{-1}) b_{Q_k}}, \tag{3.6}$$

where

$$\varepsilon \triangleq \beta [b_v + \frac{k_g(b_{x0} + b_w T)}{1 - k_f O(\lambda^{-1})}]; \tag{3.7}$$

- *2). for any prescribed tracking error tolerance ε^*, there exists a choice of learning gain matrices $Q_k(t), k = 0, 1, \cdots, N$ and a finite number of ILC iterations M such that*

$$\sup_{t \in [0,T]} \|e_i(t)\| \leq \varepsilon^*, \qquad \forall i \geq M.$$

Before presenting a proof of the above theorem, we introduce the following lemma first.

Lemma 3.3.1 *Suppose $b_D \neq O_1(\lambda^{-1})$, then, under A2), we have*

$$\lim_{b_{Q_0} \to \infty} \beta = 0. \tag{3.8}$$

Proof. Rewrite the β expression as

$$\beta = \frac{b_{Q_0}^{-1}}{\eta^{-1} b_{Q_0}^{-1} - O_1(\lambda^{-1})}. \tag{3.9}$$

Next, we show that $\eta^{-1} b_{Q_0}^{-1} - O_1(\lambda^{-1})$ tends to a finite number other than zero when b_{Q_0} approaches to infinity. It is obvious that there exists a t^* such that,

$$\|I_r + D(t)Q_0(t)\| \geq \|I_r + D(t^*)Q_0(t^*)\|, \quad \forall t \in [0, T].$$

By the fact that

$$\|(I_r + D(t)Q_0(t))^{-1}\| \geq \|I_r + D(t)Q_0(t)\|^{-1}, \quad \forall t \in [0, T],$$

we have

$$\eta^{-1}b_{Q_0}^{-1} = \frac{1}{b_{Q_0}\sup_{t\in[0,T]}\|(I_r + D(t)Q_0(t))^{-1}\|}$$

$$\leq \frac{1}{b_{Q_0}\sup_{t\in[0,T]}\|(I_r + D(t)Q_0(t))\|^{-1}}$$

$$= \frac{\|I_r + D(t^*)Q_0(t^*)\|}{b_{Q_0}}$$

$$\leq \frac{1 + b_D b_{Q_0}}{b_{Q_0}} = b_D + b_{Q_0}^{-1}$$

This implies that when $b_D \neq O_1(\lambda^{-1})$, $\lim_{b_{Q_0}\to\infty}\beta = 0$.

Now we proceed to present a proof of Theorem 3.3.1.

Proof. 1). Using (3.4), we can write

$$e_{i+1}(t) = e_i(t) - \Delta y_{i+1}(t)$$

$$= e_i(t) - D(t)\Delta u_{i+1}(t) - \Delta g_{i+1} - \Delta v_{i+1}$$

$$= e_i(t) - D(t)\sum_{k=0}^{N} Q_k(t)e_l(t) - \Delta g_{i+1} - \Delta v_{i+1}. \qquad (3.10)$$

Combining $e_{i+1}(t)$ terms in both sides of (3.10) gives

$$e_{i+1}(t) = [I_r + D(t)Q_0(t)]^{-1}\{[I_r - D(t)Q_1(t)]e_i(t)$$

$$- \sum_{k=2}^{N} D(t)Q_k(t)e_l(t) - \Delta g_{i+1} - \Delta v_{i+1}\}. \qquad (3.11)$$

Taking norm operation of (3.11) gives

$$\|e_{i+1}(t)\| \leq \eta(\sum_{k=1}^{N}\alpha_k\|e_l(t)\| + b_v + k_g\|\Delta x_{i+1}(t)\|). \qquad (3.12)$$

By integrating the state equation (3.1) and then taking its norm, we have

$$\|\Delta x_{i+1}(t)\| \leq \|\Delta x_{i+1}(0)\| + \int_0^t (\|\Delta f_{i+1}\| + \|\Delta w_{i+1}(\tau)\|)d\tau$$

$$\leq b_{x0} + b_w T + k_f \int_0^t (\|\Delta x_{i+1}(\tau)\| + \|\Delta u_{i+1}(\tau)\|)d\tau. \quad (3.13)$$

Noticing the fact that

$$\int_0^t \|\Delta x_{i+1}(\tau)\|e^{-\lambda t}d\tau \leq \|\Delta x_{i+1}(t)\|_\lambda \int_0^t e^{-\lambda(t-\tau)}d\tau$$

$$\leq \|\Delta x_{i+1}(t)\|_\lambda O(\lambda^{-1}),$$

$$(3.14)$$

multiplying $e^{-\lambda t}$ to both sides of (3.13) and taking the $\lambda-$norm operation yield

$$\|\Delta x_{i+1}(t)\|_\lambda \leq \frac{b_{x0} + b_w T + k_f O(\lambda^{-1})\|\Delta u_{i+1}(t)\|_\lambda}{1 - k_f O(\lambda^{-1})}, \tag{3.15}$$

where λ is chosen to be sufficiently large to ensure that

$$1 - k_f O(\lambda^{-1}) > 0. \tag{3.16}$$

Taking the λ−norm of (3.12) and substituting of (3.15) into (3.12) give

$$\begin{aligned}
\|e_{i+1}(t)\|_\lambda &\leq \eta\{\sum_{k=1}^{N} \alpha_k \|e_l(t)\|_\lambda + b_v \\
&\quad + k_g(b_{x0} + b_w T)/[1 - k_f O(\lambda^{-1})] + O_1(\lambda^{-1})\|\Delta u_{i+1}(t)\|_\lambda\} \\
&\leq \eta\{\sum_{k=1}^{N} \rho_k \|e_l(t)\|_\lambda + b_v \\
&\quad + k_g(b_{x0} + b_w T)/[1 - k_f O(\lambda^{-1})] \\
&\quad + O_1(\lambda^{-1})b_{Q_0}\|e_{i+1}(t)\|_\lambda\},
\end{aligned} \tag{3.17}$$

where

$$\rho_k \triangleq \alpha_k + O_1(\lambda^{-1})b_{Q_k}. \tag{3.18}$$

It is obvious that a sufficiently large λ can be chosen to simultaneously satisfy (3.16) and

$$1 - \eta O_1(\lambda^{-1})b_{Q_0} > 0. \tag{3.19}$$

By rewriting (3.17), we get

$$\|e_{i+1}(t)\|_\lambda \leq \beta \sum_{k=1}^{N} \rho_k \|e_l(t)\|_\lambda + \varepsilon, \tag{3.20}$$

where ε is given in (3.7). According to (3.5), it is possible to choose a sufficiently large λ which makes

$$\beta \sum_{k=1}^{N} \rho_k < 1. \tag{3.21}$$

From Lemma 2.3.1, the final tracking error is bounded as

$$\lim_{i\to\infty} \|e_i(t)\|_\lambda \leq \frac{\varepsilon}{1 - \beta \sum_{k=1}^{N} \rho_k} = \frac{\varepsilon}{1 - \rho - \beta \sum_{k=1}^{N} O_1(\lambda^{-1})b_{Q_k}}, \tag{3.22}$$

where ρ is given in (3.5).

2). Based on Lemma 3.3.1, $\beta \to 0$ when $b_{Q_0} \to \infty$. In this case, we know $\beta \sum_{k=1}^{N} \rho_k \to 0$ and also $\varepsilon \to 0$. Thus we can conclude that, for any prescribed tracking error tolerance ε^*, there exist a choice of learning gain matrices $Q_k(t), k = 0, 1, \cdots, N$ and a finite number of ILC iterations M such that $\sup_{t\in[0,T]} \|e_i(t)\| \leq \varepsilon^*, \quad \forall i \geq M.$

Regarding Theorem 3.3.1, we make the following remarks.

Remark 3.3.1. In the case that at every ILC iteration the uncertainty, disturbance, and initialization error are all the same, i.e., they are repetitive, the
final tracking error bound will be zero according to (3.7). We can see that
the Assumption A1) is less restrictive than the conventional boundedness
assumption given in , e.g., [215].

Remark 3.3.2. Suppose $x(t) = 0$ when $t < 0$. Then $\sup_{t \in [t_0, T]} \|x(t - \tau)\| \leq \sup_{t \in [t_0, T]} \|x(t)\|$, where $0 \leq \tau \leq T$ is the time delay in the states. Therefore,
following the proof of Theorem 3.3.1, it can be easily verified that the result
of Theorem 3.3.1 still holds for systems with unknown time-varying delays
in state variables.

Remark 3.3.3. Consider the case that the desired trajectory varies with respect to ILC iteration number i. Suppose the desired trajectory at i-th iteration is changed to $y_{d_i}(t)$. If

$$\|y_{d_{i+1}}(t) - y_{d_i}(t)\| < b_{y_d}, \quad \forall t \in [0, T] \text{ and } \forall i, \tag{3.23}$$

then, all the discussions made above are all valid by replacing b_v with $b_v + b_{y_d}$.

3.4 Simulation Results

A number of simulations were carried out to verify the results presented
in previous section. A typical simulation example is now presented in this
section. Consider a repetitive nonlinear system described by

$$\begin{cases} \begin{bmatrix} \dot{x}_{1_i}(t) \\ \dot{x}_{2_i}(t) \end{bmatrix} = \begin{bmatrix} \sin(x_{2_i}(t))u_{2_i}(t) & 1 + \sin(x_{1_i}(t))u_{1_i}(t) \\ -2 - 5t & -3 - 2t \end{bmatrix} \begin{bmatrix} x_{1_i}(t) \\ x_{2_i}(t) \end{bmatrix} + \begin{bmatrix} w_{1_i}(t) \\ w_{2_i}(t) \end{bmatrix} \\ \\ \begin{bmatrix} y_{1_i}(t) \\ y_{2_i}(t) \end{bmatrix} = \begin{bmatrix} 4x_{1_i}(t)\sin(x_{2_i}(t)) \\ x_{2_i}(t)\cos(x_{1_i}(t)) \end{bmatrix} + \begin{bmatrix} \cos(10\pi t) & -\sin(10\pi t) \\ \sin(10\pi t) & \cos(10\pi t) \end{bmatrix} \begin{bmatrix} u_{1_i}(t) \\ u_{2_i}(t) \end{bmatrix} \\ \qquad\qquad + \begin{bmatrix} v_{1_i}(t) \\ v_{2_i}(t) \end{bmatrix} \end{cases}$$

where i is the system repetition number and the time $t \in [0, 1]$. The uncertainty and output disturbance varying with respect to both time t and ILC
iteration number i are

$$\begin{bmatrix} w_{1_i}(t) \\ w_{2_i}(t) \end{bmatrix} \triangleq i\gamma \begin{bmatrix} \cos(2\pi f_0 t) \\ 2\cos(4\pi f_0 t) \end{bmatrix}, \quad \begin{bmatrix} v_{1_i}(t) \\ v_{2_i}(t) \end{bmatrix} \triangleq i\gamma \begin{bmatrix} \sin(2\pi f_0 t) \\ 2\sin(4\pi f_0 t) \end{bmatrix},$$

where $\gamma = 0.01$, $f_0 = 1/(20h)$ Hertz and h is the time step in the simulation
for integration. The desired tracking trajectories are

$$y_{1_d}(t) = y_{2_d}(t) \triangleq 12t^2(1 - t).$$

It is easy to see that the desired initial states are $x_{1_d}(0) = x_{2_d}(0) = 0$ and $D^{-1}(t) = D^T(t)$. We use the second order ILC updating law ($N = 2$) and assume $x_{1_i}(0) = x_{2_i}(0) = 0$. Let the learning gain matrices be given by $Q_k(t) = \alpha'_k D^T(t), k = 0, 1, 2$, where α'_0, α'_1, and α'_2 are constants. Denote $\alpha' = [\alpha'_0, \alpha'_1, \alpha'_2]$. The following cases have been simulated.

Case 1 : $\alpha' = [0.0,\ 0.5,\ 0.0]$; Case 4 : $\alpha' = [0.0,\ 0.5,\ 0.2]$;
Case 2 : $\alpha' = [0.1,\ 0.5,\ 0.0]$; Case 5 : $\alpha' = [0.1,\ 0.5,\ 0.2]$;
Case 3 : $\alpha' = [0.5,\ 0.5,\ 0.0]$; Case 6 : $\alpha' = [0.5,\ 0.5,\ 0.2]$.

The RK-4 method is used to numerically integrate the state equation with a constant time step $h = 0.01$ second. For comparison, the maximum iteration number are set to be 15 and the tracking error tolerance ε^* to be 0.02 for all cases. The following infinite norm of the tracking error (error bound) is used.

$$b_{e_j}(i) \overset{\triangle}{=} \sup_{t \in [0,T]} \mid e_{j_i}(t) \mid, \quad j = 1, 2.$$

The iteration histories of b_{e_1} and b_{e_2} are compared in Fig. 3.1 for Cases 1, 2, and 3 where the first order ILC updating law ($N = 1$) is used with the same learning gain $Q_1(t)$. By increasing the value of the learning gain factor α'_0 of the CITE from 0 to 0.5, a faster convergence as well as a smaller tracking error bound at each ILC iteration can be observed. When α'_0 is 0.5, the prescribed tracking accuracy ε^* is satisfied. The same observation can be made from Fig. 3.2 for cases 4, 5, and 6 where the order of the ILC updating law $N = 2$ with the same learning gains $Q_1(t)$ and $Q_2(t)$.

By the comparison between Figs. 3.1 and 3.2, we can see that with a high-order ILC algorithm, better transient behavior along iteration number i direction can be achieved. Thus we believe that the high-order terms in ILC updating law can improve the transient performance of ILC process. This is also supported by Fig. 3.3 which gives a comparison of iteration histories of root-mean-squares (RMS) values of the tracking errors for Cases 1 ($N = 1$) and Case 5 (N=2). Regarding the adjustment of the transient behavior of ILC process, [139] showed in detail the role of the P-component in a D-type ILC updating law.

3.5 Conclusion

A high-order P-type updating law utilizing the current iteration tracking error for iterative learning control of a class of uncertain nonlinear systems is proposed. An ILC convergence condition is established. It is shown that the tracking error is uniformly bounded in the presence of uncertainty, disturbance, and initialization error. It is shown that the final tracking error bound is directly dependent on the bounds of the differences between two successive ILC iterations of the uncertainty, the disturbance, and the initialization error. The most attractive feature of the proposed ILC algorithm is that the

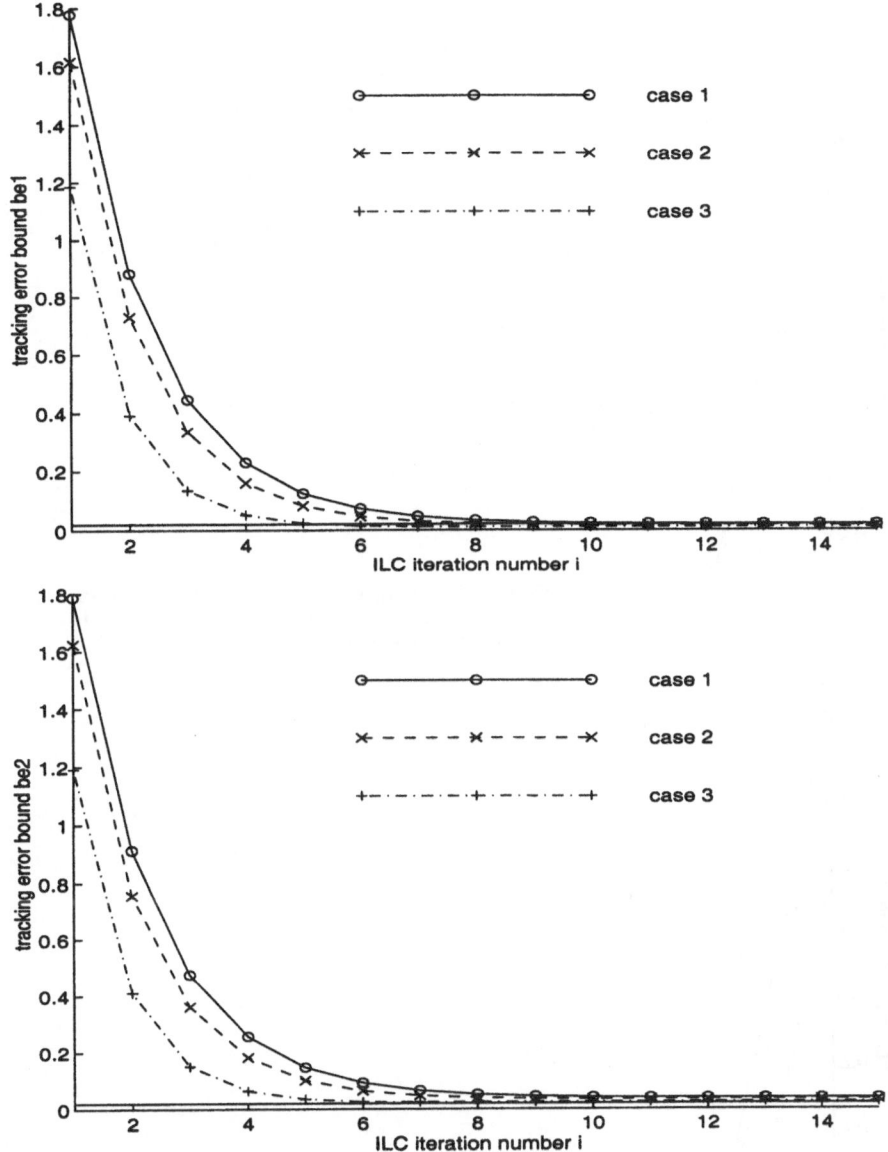

Fig. 3.1. The effect of current iteration tracking error in ILC updating law (N=1)

size of the final tracking error bound and the rate of ILC convergence can be adjusted by learning gain of current iteration tracking error, which makes the proposed scheme more attractive from the viewpoint of practical application.

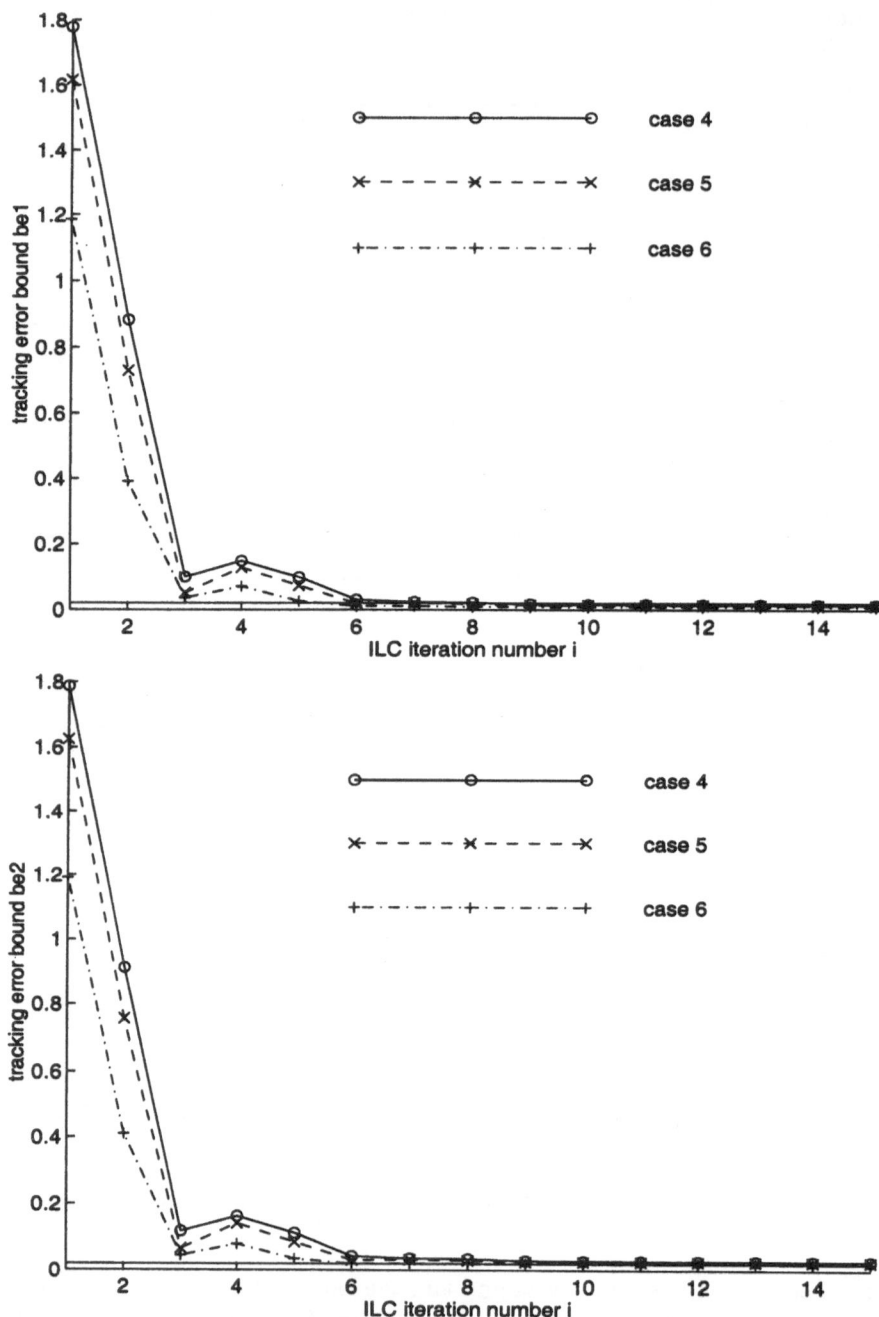

Fig. 3.2. The effect of current iteration tracking error in ILC updating law (N=2)

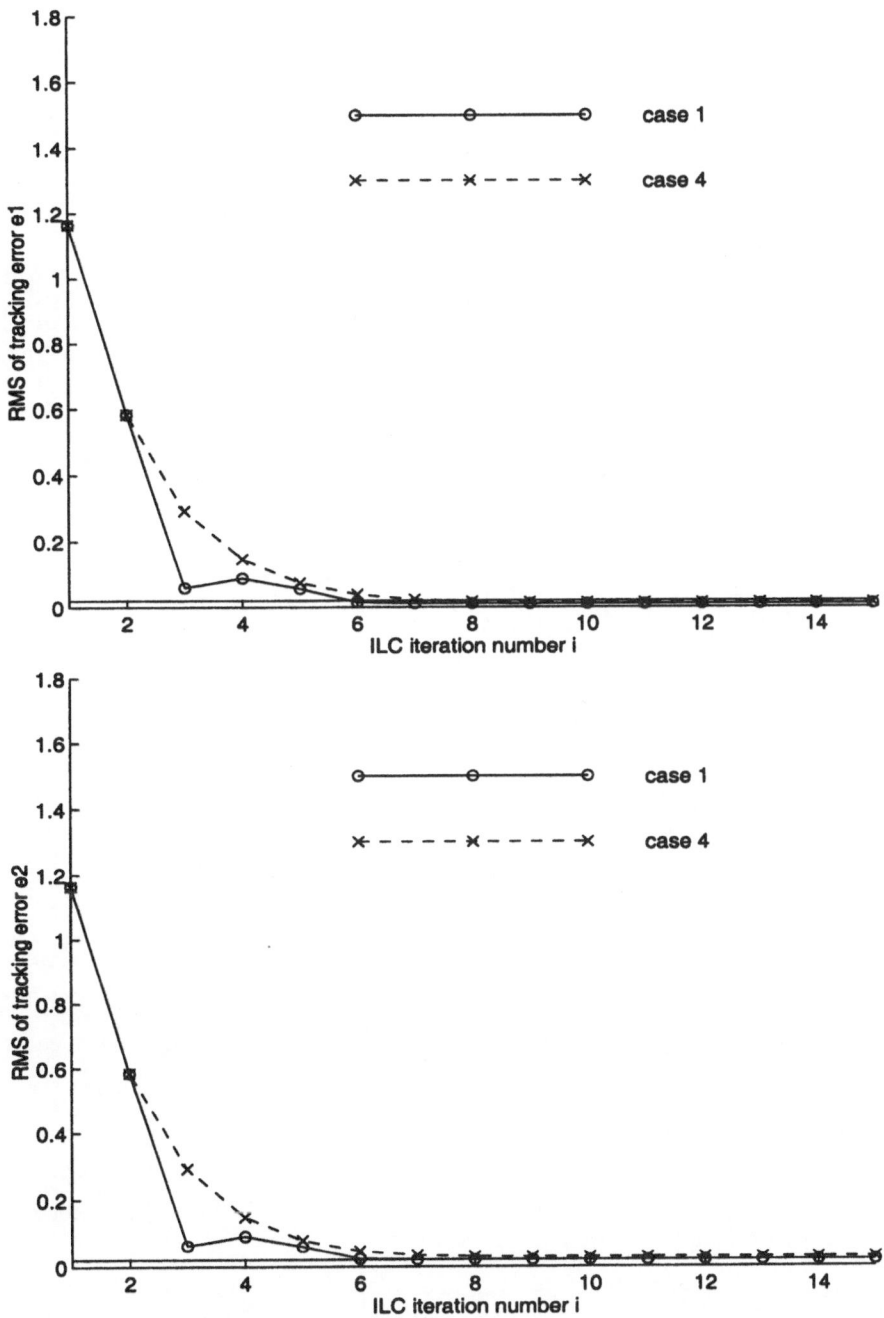

Fig. 3.3. The comparison of ILC updating laws with $N = 1$ and $N = 2$ ($\alpha_0' = 0.1$)

4. Iterative Learning Control for Uncertain Nonlinear Discrete-time Systems Using Current Iteration Tracking Error

4.1 Introduction

ITERATIVE LEARNING CONTROL is essentially a memory-based scheme. For the implementation of ILC algorithms, a discrete-time form of the algorithm is more convenient. Thus its theoretical analysis is also important. Some results were obtained for linear discrete-time systems by using 2-D system theory [89, 136, 216], or by including a parameter estimator [114], by applying the approximated impulse sequence [118, 232] and by Taylor series expansion [238]. Nonlinear discrete-time systems were also considered in [114, 238, 119]. But the robustness issue has not been well discussed except some analysis in Saab [217]. The result is parallel to that of [103]. Generally, robustness analysis is to show the boundedness of the tracking errors in the presence of the bounded uncertainty, disturbance and re-initialization error.

It has been shown in Chapter 3 that the CITE information is helpful in improving the ILC convergence properties for continuous-time nonlinear systems. In this chapter, we shall explicitly explain the CITE role in the iterative learning control for discrete-time nonlinear systems. Under some relaxed conditions, similar conclusions as in Chapter 3 can be drawn, i.e., the tracking error bound and the ILC convergence rate can be tuned by the learning gain of the current iteration tracking error in the ILC updating law if a suitable prediction scheme is applied. The effectiveness of the proposed iterative learning controller is illustrated by simulation.

This chapter is organized as follows. The CITE-assisted discrete-time ILC problem is formulated in Sec. 4.2 where some of the notations and preliminaries are introduced. Sec. 4.3 details the robust convergence analysis for the proposed ILC schemes. An illustrative simulation is presented in Sec. 4.4. Finally, the conclusions are drawn in Sec. 4.5.

4.2 Problem Setting and Preliminaries

Consider a discrete-time uncertain nonlinear time-varying system performing repetitive tasks as follows

$$\begin{cases} x_i(t+1) = f(x_i(t),t) + B(x_i(t),t)u_i(t) + w_i(t) \\ y_i(t) = C(t)x_i(t) + v_i(t) \end{cases} \tag{4.1}$$

where i denotes the i-th repetitive operation of the system; t is the discrete time index and $t \in [0, N]$ which means that $t \in \{0, 1, \cdots, N\}$; $x_i(t) \in R^n$, $u_i(t) \in R^m$, and $y_i(t) \in R^r$ are the state, control input, and output of the system, respectively; $C(t) \in R^{r \times n}$ is time-varying matrix; the functions $f(\cdot, \cdot) : R^n \times [0, N] \mapsto R^n$ and $B(\cdot, \cdot) : R^n \times [0, N] \mapsto R^m$ are uniformly globally Lipschitzian in x, i.e., $\forall t \in [0, N], \forall i, \exists$ constants k_f, k_B, such that

$$\|\Delta f_i(t)\| \leq k_f \|\Delta x_i(t)\|, \quad \|\Delta B_i(t)\| \leq k_B \|\Delta x_i(t)\|$$

where $\Delta f_i(t) \triangleq f(x_i(t), t) - f(x_{i-1}(t), t)$, $\Delta B_i(t) \triangleq B(x_i(t), t) - B(x_{i-1}(t), t)$, $\Delta x_i(t) \triangleq x_i(t) - x_{i-1}(t)$; $w_i(t)$, $v_i(t)$ are uncertainty or disturbance to the system bounded with unknown bounds b_w, b_v defined as

$$b_w \triangleq \sup_{t \in [0,N]} \|w_i(t)\|, \quad b_v \triangleq \sup_{t \in [0,N]} \|v_i(t)\|, \; \forall i. \tag{4.2}$$

Denote the output tracking error $e_i(t) \triangleq y_d(t) - y_i(t)$ where $y_d(t)$ is the given desired output trajectory, which is realizable, i.e., given a bounded $y_d(t)$, there exists a unique bounded desired input $u_d(t)$, $t \in [0, N]$ such that when $u(t) = u_d(t)$, the system has a unique bounded desired state $x_d(t)$ satisfying

$$\begin{cases} x_d(t+1) = f(x_d(t), t) + B(x_d(t), t)u_d(t) \triangleq f_d + B_d u_d \\ y_d(t) = C(t)x_d(t) \triangleq C(t)x_d. \end{cases} \tag{4.3}$$

Denote the bound of the desired control u_d as $b_{u_d} \triangleq \sup_{t \in [0,N]} \|u_d(t)\|$. Then the problem is formulated as follows. Starting from an arbitrary initial control input $u_0(t)$, obtain the next control input $u_1(t)$ and the subsequent series $\{u_i(t) \mid i = 2, 3, \cdots\}$ for system (4.1) by using a proper learning control updating law in such a way that when $i \to \infty, y_i(t) \to y_d(t)$. In the presence of uncertainty, disturbance and initialization error, it is desired that the tracking error bound $y_i(t) - y_d(t)$ as well as the ILC convergence rate can be adjusted to a prescribed level.

To solve the above problem, we propose a simple ILC updating law which includes a CITE term, i.e.,

$$u_{i+1}(t) = u_i(t) + Q_0(t)e_{i+1}(t+1) + Q_1(t)e_i(t+1) \tag{4.4}$$

where $Q_j(t) \in R^{m \times r}(j = 0, 1)$ are the learning matrices.

Remark 4.2.1. From the causality consideration in the discrete-time domain, one may use a proper extrapolation (static) or filtering/prediction (dynamic) scheme to obtain the tracking error at instant $t + 1$. This is illustrated by simulation results. Moreover, ILC scheme (4.4) is essentially a PD-type one because, in fact, suppose the sampling time is T_s, then $e_i(t+1) \approx e_i(t) + T_s \dot{e}_i(t)$.

To restrict our discussion, the following assumptions are made.

- A1). The initialization error is bounded as follows, $\forall t \in [0, N]$, $\forall i$, $\|x_d(0) - x_i(0)\| \leq b_{x_0}$, $\|y_d(0) - y_i(0)\| \leq b_C b_{x_0} + b_v$, where $b_C \overset{\triangle}{=} \sup_{t \in [0,N]} \|C(t)\|$.
- A2). Matrix $C(t+1)B(x(t), t)$ has a full column rank for all $t \in [0, N]$, $x(t) \in R^n$.
- A3). Operator $B(\cdot, \cdot)$ is bounded, i.e., $\exists$ a constant b_B such that for all i, $\sup_{t \in [0,N]} \|B(x_i(t), t)\| \overset{\triangle}{=} \sup_{t \in [0,N]} \|B_i(t)\| \leq b_B$.

Assumption A1) restricts that the initial states or the initial outputs in each repetitive operation should be inside a given ball centered at the desired initial ones. The radius of the ball may be unknown. The number of outputs r must be less than or equal to the number of inputs m according to A2). A3) says that the range of operator $B(\cdot, \cdot)$ is always finite. This is reasonable because the repetitive tasks are performed in a finite time interval $[0, NT_s]$.

To analyze the robust convergence property of the proposed CITE assisted ILC algorithm, the following $\lambda-$norm is introduced for a discrete-time vector $h(t), t = 0, 1, \cdots, N$.

$$\|h(t)\|_\lambda \overset{\triangle}{=} \sup_{t \in [0,N]} \hat{e}^{-\lambda t} \|h(t)\| \tag{4.5}$$

where $\lambda > 0$ when $\hat{e} > 1$ or $\lambda < 0$ when $\hat{e} \in (0, 1)$. In this Chapter, $\hat{e} \overset{\triangle}{=} k_f + b_{u_d} k_B$. It should be pointed out that the $\lambda-$norm used in this chapter is equivalent to the infinity-norm [217]. Also, it is noted that as $\hat{e}$ is related to the *upper* bounds of system and task-related parameters such as k_f, b_{u_d} and k_B, it is without loss of generality to assume $\hat{e} > 1$.

To facilitate the later derivations, some basic relations are presented in the following. First, it is an easy exercise to verify the following recursion formula.

$$z_{i+1} = a_1 z_i + a_2 z_i^* + a_3 = a_1^{i+1} z_0 + \sum_{j=0}^{i} a_1^{i-j}(a_2 z_j^* + a_3) \tag{4.6}$$

where $\{z_i, z_i^* \mid i = 0, 1, \cdots\}$ are two series and related each other by coefficients a_1, a_2, a_3. Let

$$\delta x_i(t) \overset{\triangle}{=} x_d(t) - x_i(t), \quad \delta u_i(t) \overset{\triangle}{=} u_d(t) - u_i(t),$$

$$\delta f_i(t) \overset{\triangle}{=} f_d - f(x_i(t), t), \quad \delta B_i(t) \overset{\triangle}{=} B_d - B_i(t).$$

Then, from (4.1) and (4.3), it can be obtained that

$$\delta x_i(t+1) = \delta f_i(t) + \delta B_i(t)u_d + B_i(t)\delta u_i(t) - w_i(t). \tag{4.7}$$

Taking the norm for (4.7) yields

$$\|\delta x_i(t+1)\| \leq \hat{e}\|\delta x_i(t)\| + b_B\|\delta u_i(t)\| + b_w. \tag{4.8}$$

Applying (4.6), we can get

$$\|\delta x_i(t+1)\| \le \hat{e}^{t+1} b_{x_0} + \sum_{j=0}^{t} \hat{e}^{t-j} (b_B \|\delta u_i(j)\| + b_w). \qquad (4.9)$$

To see a simple relationship between $\|\delta x_i(t)\|_\lambda$ and $\|\delta u_i(t)\|_\lambda$, by noticing the following facts that

- $\|c\|_\lambda \equiv | c |, \forall c \in R$;
- $\forall | \lambda |> 1, \sup_{t\in[0,N]} \hat{e}^{-(\lambda-1)t} = 1$;
- $\forall t_1 \in [0, N_1], t_2 \in [0, N_2]$, if $0 \le N_1 \le N_2 \le N$, then $\|\delta h(t_1)\|_\lambda \le \|\delta h(t_2)\|_\lambda$;

then, taking the $\lambda-$norm ($| \lambda |> 1$) operation of (4.9) gives

$$\|\delta x_i(t)\|_\lambda \le b_{x_0} + b_B O(| \lambda |^{-1})\|\delta u_i(t)\|_\lambda + c_0 b_w \qquad (4.10)$$

where

$$O(| \lambda |^{-1}) \triangleq \frac{1 - \hat{e}^{-(\lambda-1)N}}{\hat{e}^\lambda - \hat{e}}, \quad c_0 \triangleq \sup_{t\in[0,N]} \frac{\hat{e}^{-(\lambda-1)t}(1 - \hat{e}^{-t})}{\hat{e} - 1}.$$

For brevity of our discussion, in the sequel, the following notations are used.

$$b_{Q_0} \triangleq \sup_{t\in[0,N]} \|Q_0(t)\|, \quad b_{Q_1} \triangleq \sup_{t\in[0,N]} \|Q_1(t)\|,$$

$$\rho \triangleq \sup_{t\in[0,N]} \|I_m - Q_1(t)C(t+1)B_i(t)\|, \quad \forall i,$$

$$\eta \triangleq \sup_{t\in[0,N]} \|(I_m + Q_0(t)C(t+1)B_i(t))^{-1}\|, \quad \forall i,$$

$$O_0(| \lambda |^{-1}) \triangleq b_{Q_0} b_C \hat{e} b_B O(| \lambda |^{-1}),$$

$$O_1(| \lambda |^{-1}) \triangleq b_{Q_1} b_C \hat{e} b_B O(| \lambda |^{-1}),$$

$$\beta \triangleq \eta/[1 - \eta O_0(| \lambda |^{-1})].$$

4.3 Main Results and Convergence Analysis

A main result is presented in the following theorem.

Theorem 4.3.1 *Consider the repetitive discrete-time uncertain time-varying nonlinear system (4.1) under assumptions A1)-A3). Given the realizable desired trajectory $y_d(t)$ over the fixed time interval $[0, NT_s]$ and using the ILC updating law (4.4), if*

$$\beta\rho < 1, \qquad (4.11)$$

is satisfied, then the λ-norm of the tracking errors $e_i(t)$, $\delta u_i(t)$, $\delta x_i(t)$ are all bounded. For a sufficiently large $| \lambda |$, $\forall t \in [0, N]$,

$$b_u \overset{\triangle}{=} \lim_{i \to \infty} \|\delta u_i(t)\|_\lambda \le \frac{\beta \varepsilon}{1 - \beta \bar{\rho}}, \tag{4.12}$$

where

$$\bar{\rho} \overset{\triangle}{=} \rho + O_1(|\lambda|^{-1}),$$

$$\varepsilon \overset{\triangle}{=} (b_{Q_0} + b_{Q_1})\varepsilon_0, \tag{4.13}$$

$$\varepsilon_0 \overset{\triangle}{=} b_C \hat{e}(b_{x_0} + c_0 b_w) + b_C b_w + b_v. \tag{4.14}$$

Also, we have

$$b_x \overset{\triangle}{=} \lim_{i \to \infty} \|\delta x_i(t)\|_\lambda \le b_{x_0} + b_B O(|\lambda|^{-1})b_u + c_0 b_w, \tag{4.15}$$

$$b_e \overset{\triangle}{=} \lim_{i \to \infty} \|e_i(t)\|_\lambda \le b_C b_x + b_v. \tag{4.16}$$

Moreover, b_u, b_x, b_e will all converge uniformly to zero for $t = 0, 1, \cdots, N$ as $i \to \infty$ in absence of uncertainty, disturbance and initialization error, i.e., $b_w, b_v, b_{x_0} \to 0$.

Before presenting a proof of the above theorem, we introduce the following lemma first.

Lemma 4.3.1 *From the above notations, we have*

$$\lim_{b_{Q_0} \to \infty} \beta = 0. \tag{4.17}$$

Proof. Rewrite the expression of β as

$$\beta = \frac{b_{Q_0}^{-1}}{\eta^{-1} b_{Q_0}^{-1} - b_{Q_0}^{-1} O_0(|\lambda|^{-1})}. \tag{4.18}$$

It's enough to show that $\eta^{-1} b_{Q_0}^{-1} - b_{Q_0}^{-1} O_0(|\lambda|^{-1})$ tends to a finite number other than zero when b_{Q_0} approaches to infinity. Because

$$b_{Q_0}^{-1} O_0(|\lambda|^{-1}) = b_C \hat{e} b_B O(|\lambda|^{-1})$$

therefore, we can write β as

$$\beta \le \frac{b_{Q_0}^{-1}}{\eta^{-1} b_{Q_0}^{-1} - b_C \hat{e} b_B O(|\lambda|^{-1})}.$$

By the fact that $\forall t \in [0, N], \forall i$,

$$\|(I_m + Q_0(t)C(t+1)B_i(t))^{-1}\| \ge \|I_m + Q_0(t)C(t+1)B_i(t)\|^{-1},$$

and because it is obvious that there exists a t^* such that, $\forall t \in [0, N]$,

$$\|I_m + Q_0(t)C(t+1)B_i(t)\| \ge \|I_m + Q_0(t^*)C(t^*+1)B_i(t^*)\|,$$

clearly, we have

$$\eta^{-1}b_{Q_0}^{-1} = \frac{1}{b_{Q_0}\sup_{t\in[0,N]}\|(I_m + Q_0(t)C(t+1)B_i(t))^{-1}\|}$$

$$\leq \frac{1}{b_{Q_0}\sup_{t\in[0,N]}\|I_m + Q_0(t)C(t+1)B(t)\|^{-1}}$$

$$= \frac{\|I_m + Q_0(t^*)C(t^*+1)B(t^*)\|}{b_{Q_0}}$$

$$\leq \frac{1 + b_C b_B b_{Q_0}}{b_{Q_0}} = b_C b_B + b_{Q_0}^{-1}. \tag{4.19}$$

As there is no restriction on $|\lambda|$, it can be guaranteed that $\hat{e}O(|\lambda|^{-1}) \neq 1$, which implies that $\lim_{b_{Q_0}\to\infty}\beta = 0$.

Remark 4.3.1. From Lemma 4.3.1, it is obvious the convergence condition (4.11) for the CITE assisted ILC algorithm is relaxed compared to the convergence condition given in [217].

Now we proceed to prove Theorem 4.3.1.

Proof. The tracking error at the $(i+1)$−th repetition is

$$e_{i+1}(t) = y_d(t) - y_{i+1}(t)$$
$$= C(t)\delta x_{i+1}(t) - v_{i+1}(t) \tag{4.20}$$

Investigating the control deviation at the $(i+1)$-th repetition $\delta u_{i+1}(t)$ gives

$$\delta u_{i+1}(t) = \delta u_i(t) - Q_0(t)e_{i+1}(t+1) - Q_1(t)e_i(t+1)$$
$$= \delta u_i(t) - Q_0(t)C(t+1)\delta x_{i+1}(t+1)$$
$$\quad - Q_1(t)C(t+1)\delta x_i(t+1)$$
$$\quad + Q_0(t)v_{i+1}(t+1) + Q_1(t)v_i(t+1) \tag{4.21}$$

By referring to (4.7), (4.21) can be written as

$$\delta u_{i+1}(t) = \delta u_i(t) - Q_0(t)C(t+1)[\delta f_{i+1}(t)$$
$$+ \delta B_{i+1}(t)u_d + B_{i+1}(t)\delta u_{i+1}(t) - w_{i+1}(t)]$$
$$- Q_1(t)C(t+1)[\delta f_i(t) + \delta B_i(t)u_d + B_i(t)\delta u_i(t)$$
$$- w_i(t)] + Q_0(t)v_{i+1}(t+1) + Q_1(t)v_i(t+1). \tag{4.22}$$

Collecting terms and then performing the norm operation for (4.22) yield

$$\|\delta u_{i+1}(t)\| \leq \eta[\rho\|\delta u_i(t)\| + b_C\hat{e}(b_{Q_0}\|\delta x_{i+1}(t)\| + b_{Q_1}\|\delta x_i(t)\|)$$
$$+ (b_{Q_0} + b_{Q_1})(b_C b_w + b_v)]. \tag{4.23}$$

By utilizing the relationship in (4.10), taking the λ−norm for (4.23) gives

$$\|\delta u_{i+1}(t)\|_\lambda \leq \eta[\rho\|\delta u_i(t)\|_\lambda + O_0(|\lambda|^{-1})\|\delta u_{i+1}(t)\|_\lambda$$
$$+ O_1(|\lambda|^{-1})\|\delta u_i(t)\|_\lambda + \varepsilon]. \tag{4.24}$$

Because it is clear that a sufficiently large $|\lambda|$ can be used to ensure that

$$\eta O_0(|\lambda|^{-1})\| < 1, \tag{4.25}$$

therefore, (4.24) can be written as

$$\|\delta u_{i+1}(t)\|_\lambda \le \beta(\bar{\rho}\|\delta u_i(t)\|_\lambda + \varepsilon). \tag{4.26}$$

Obviously, there exists a sufficiently large $|\lambda|$ that satisfies (4.25) and the condition $\beta\bar{\rho} < 1$ simultaneously. It is easy to see

$$\lim_{i \to \infty} \|\delta u_i(t)\|_\lambda = \frac{\beta\varepsilon}{1 - \beta\bar{\rho}}. \tag{4.27}$$

From (4.10) and (4.20), (4.15) and (4.16) can be verified. Moreover, b_u, b_x, b_e will all converge uniformly to zero for $t = 0, 1, \cdots, N$ as $i \to \infty$ in absence of uncertainty, disturbance and initialization error, i.e., $b_w, b_v, b_{x_0} \to 0$.

Remark 4.3.2. According to Lemma 4.3.1, it can be observed that the rate of convergence $\beta\bar{\rho}$ and final tracking error bound can be tuned by adjusting b_{Q_0}.

Remark 4.3.3. As acclaimed in [17], the convergence condition (4.11) does not mean that a full knowledge of the dynamic system (4.1) is necessary. As it will be shown in the simulation, the knowledge of the system model appeared in (4.11) may not be necessarily known or accurately known.

Remark 4.3.4. Consider the case that the desired trajectory varies with respect to ILC iteration number i. Suppose the desired trajectory at i-th iteration is changed to $y_{d_i}(t)$. If

$$\|y_{d_i}(t) - y_d(t)\| < b_{y_d}, \quad \forall t \in [0, N] \text{ and } \forall i, \tag{4.28}$$

then, all the discussions made above are all valid by replacing b_v with $b_v + b_{y_d}$.

Corollary 4.3.1 *If the ILC updating law (4.4) is revised as*

$$u_{i+1}(t) = u_i(t) + Q_0(t)e_{i+1}(t+1) + Q_1(t)e_i(t+1) + Q_2(t)e_i(t), \tag{4.29}$$

where the learning matrix $Q_2(t)$ is with bound denoted by b_{Q_2}, then, the results of Theorem 4.3.1 still hold with modified ε and $\bar{\rho}_1$.

Including $e_i(t)$ in the ILC updating law (4.4) may improve the ILC convergence property. Considering the case when $Q_2(t) = -Q_1(t)$, it is clearly shown that this is an equivalence of the conventional D-type ILC in the continuous-time situation.

Proof. Similar to proof of Theorem 4.3.1. The result follows the change that an extra-term $-Q_2(t)e_i(t)$ is added to (4.21) and (4.22). Then the similar convergence results as in Theorem 4.3.1 come from the revised form of (4.23):

$$\begin{aligned}
\|\delta u_{i+1}(t)\| \le &\, \eta[\rho\|\delta u_i(t)\| \\
&+ [(b_{Q_0} + b_{Q_1})\hat{e} + b_{Q_2}b_C]\|\delta x_i(t)\| \\
&+ (b_{Q_0} + b_{Q_1})(b_C b_w + b_v)].
\end{aligned} \tag{4.30}$$

It should be pointed out that the role of b_{Q_0} is mainly for tuning the rate of ILC convergence according to Theorem 4.3.1. The convergence condition (4.11) is more easily satisfied by the property of β presented in Lemma 4.3.1. However, referring to (4.12)-(4.16), the adjustment of the final tracking error bound by b_{Q_0} is not desirable. Actually, $\beta\varepsilon$ in (4.12) will vary from $b_{Q_1}\varepsilon_0$ to $\varepsilon_0/\{b_C b_B[1 - \hat{e}O(|\lambda|^{-1})]\}$ when b_{Q_0} increases from 0 to ∞. It is desired to tune the b_e to a predetermined level by increasing b_{Q_0}. In the following, it will be shown that this can be achieved under more restrictive conditions on the structure of the system dynamics.

Assume that

- A4). $\forall i$, $\forall t \in [0, N]$, $(I_r + C(t + 1)B_i(t)Q_0(t))^{-1}$ always exists.

According to Theorem 4.3.1, the control is bounded at each ILC iteration. Let

$$b_{u^*} \triangleq \sup_{t \in [0,T]} \|u_i(t)\|, \forall i. \tag{4.31}$$

In the following, a similar λ−norm definition as in (4.5) is used with $\hat{e}$ replaced by

$$\tilde{e} \triangleq k_f + b_{u^*} k_B. \tag{4.32}$$

Denote $\Delta h_i(t) \triangleq h_i(t) - h_{i-1}(t)$, $h \in \{u, y, w, v\}$. Similar to (4.7) and (4.8), the following two inequalities can be obtained

$$\Delta x_i(t + 1) = \Delta f_i(t) + \Delta B_i(t)u_i(t) + B_i(t)\Delta u_i(t) + \Delta w_i(t), \tag{4.33}$$

$$\|\Delta x_i(t + 1)\| \leq \tilde{e}\|\Delta x_i(t)\| + b_B\|\Delta u_i(t)\| + b'_w. \tag{4.34}$$

where $b'_w \triangleq \sup_{t \in [0,N]} \|\Delta w_i(t)\|$, $\forall i$. It is also apparent by referring to (4.10) to have

$$\|\Delta x_i(t)\|_\lambda \leq b'_{x_0} + b_B\tilde{O}(|\lambda|^{-1})\|\Delta u_i(t)\|_\lambda + c'_0 b'_w \tag{4.35}$$

where $b'_{x_0} \triangleq \|\Delta x_i(0)\|$, $\forall i$, and

$$\tilde{O}(|\lambda|^{-1}) \triangleq \frac{1 - \tilde{e}^{-(\lambda-1)N}}{\tilde{e}^\lambda - \tilde{e}}, \quad c'_0 \triangleq \sup_{t \in [0,N]} \frac{\tilde{e}^{-(\lambda-1)t}(1 - \tilde{e}^{-t})}{\tilde{e} - 1}.$$

Similarly, the following notations are employed.

$$\rho' \triangleq \sup_{t \in [0,N]} \|I_r - C(t + 1)B_i(t)Q_1(t)\|, \quad \forall i,$$

$$\eta' \triangleq \sup_{t \in [0,N]} \|(I_m + C(t + 1)B_i(t)Q_0(t))^{-1}\|, \quad \forall i,$$

$$\tilde{O}_0(|\lambda|^{-1}) \triangleq b_{Q_0} b_C \tilde{e} b_B O(|\lambda|^{-1}),$$

$$\tilde{O}_1(|\lambda|^{-1}) \triangleq b_{Q_1} b_C \bar{e} b_B \bar{O}(|\lambda|^{-1}),$$

$$\beta' \triangleq \eta'/[1 - \eta'\tilde{O}_0(|\lambda|^{-1})], \quad b_v' \triangleq \sup_{t \in [0,N]} \|\Delta v_i(t)\|, \quad \forall i.$$

Now it is prepared to present an attractive property of the tracking error bound that it can be arbitrarily adjustable.

Theorem 4.3.2 *Consider the repetitive discrete-time uncertain time-varying nonlinear system (4.1) under assumptions A1)-A4). Given the realizable desired trajectory $y_d(t)$ over the fixed time interval $[0, NT_s]$ and using the ILC updating law (4.4), the λ-norm of the tracking errors $e_i(t)$ can be tuned to a prescribed level if*

$$\beta'\rho' < 1. \tag{4.36}$$

More specifically, for a sufficiently large $|\lambda|$, $\forall t \in [1, N]$, the final tracking error bound

$$b_e' \triangleq \lim_{i \to \infty} \|\delta e_i(t)\|_\lambda \leq \frac{\beta'\varepsilon'}{1 - \beta'\bar{\rho}'}, \tag{4.37}$$

where

$$\bar{\rho}' \triangleq \rho' + \tilde{O}_1(|\lambda|^{-1}),$$

$$\varepsilon' \triangleq b_C \tilde{e}(b_{x_0}' + c_0' b_w') + b_C b_w' + b_v'. \tag{4.38}$$

It should be noted that β' has the same property as β shown in Lemma 4.3.1. The proof of Theorem 4.3.2 is in a similar fashion as that of Theorem 4.3.1.

Proof. The tracking error at $(i+1)$-th ILC iteration can be expressed as

$$e_{i+1}(t+1) = e_i(t+1) - \Delta y_{i+1}(t+1)$$
$$= e_i(t+1) - C(t+1)\Delta x_{i+1}(t+1) - \Delta v_{i+1}(t+1). \tag{4.39}$$

Substituting (4.33) and (4.4) into (4.39) gives

$$e_{i+1}(t+1) = e_i(t+1) - C(t+1)[\Delta f_{i+1}(t) +$$
$$\Delta B_{i+1}(t)u_{i+1}(t)] - C(t+1)B_i(t)[Q_0(t)e_{i+1}(t+1) +$$
$$Q_1(t)e_i(t+1)] - C(t+1)\Delta w_{i+1}(t) - \Delta v_{i+1}(t+1). \tag{4.40}$$

By collecting terms and then taking the norm for (4.40) yield

$$\|e_{i+1}(t+1)\| \leq \eta'[\rho'\|e_i(t+1)\| + b_C\tilde{e}\|\Delta x_{i+1}(t)\|$$
$$+ b_C b_w' + b_v']. \tag{4.41}$$

Taking the λ-norm for (4.41) and using the relationship (4.35), we simply have

$$\|e_{i+1}(t+1)\|_\lambda \le \eta'[\rho'\|e_i(t+1)\|_\lambda + \varepsilon'$$
$$+\tilde{O}_0(|\,\lambda\,|^{-1})\|e_{i+1}(t+1)\|_\lambda$$
$$+\tilde{O}_1(|\,\lambda\,|^{-1})\|e_i(t+1)\|_\lambda]. \tag{4.42}$$

Clearly, a sufficiently large $|\,\lambda\,|$ can be used to ensure that

$$\eta'\tilde{O}_0(|\,\lambda\,|^{-1})\| < 1, \quad \beta'\bar{\rho}' < 1. \tag{4.43}$$

Therefore, (4.42) can be written as

$$\|e_{i+1}(t+1)\|_\lambda \le \beta'(\bar{\rho}'\|e_i(t+1)\|_\lambda + \varepsilon'). \tag{4.44}$$

It is easy to verify that (4.37) holds. Furthermore, by observation of β' property and ε' expression (4.38), the final tracking error bound can be tuned to a desired level by adjusting b_{Q_0}.

Remark 4.3.5. It should be noted that, under A4), the convergence condition (4.36) can be satisfied even when $C(\cdot)B(\cdot,\cdot)$ is not full row rank as required in [136]. Moreover, the b'_{x_0}, b'_w, b'_v appeared in (4.38) imply that the tracking error bound is only affected by the bounds of the differences of uncertainty, disturbance and initialization error between successive ILC iterations. This is a relaxed requirement, too.

Corollary 4.3.2 *If the ILC updating law (4.29) is applied instead of (4.4), the results in Theorem 4.3.2 hold in a similar way.*

Proof. The proof is similar to the most parts of that for Theorem 4.3.2. Referring to (4.44), it is easy to show that we finally come with

$$\|e_{i+1}(t+1)\|_\lambda \le \beta'(\bar{\rho}'\|e_i(t+1)\|_\lambda + b_C b_B b_{Q_2}\|e_i(t)\|_\lambda + \varepsilon'). \tag{4.45}$$

However, the following arguments are in an *enumerative* way. First, **for** $t = 0$, we know from assumption A1) that $\|e_i(0)\|_\lambda \le b_C b_{x_0} + b_v$. **For** $t = 1$, we have

$$\|e_{i+1}(1)\|_\lambda \le \beta'(\bar{\rho}'\|e_i(1)\|_\lambda + b_C b_B b_{Q_2}\|e_i(0)\|_\lambda + \varepsilon'). \tag{4.46}$$

based on (4.45). Obviously, we know that $\|e_{i+1}(1)\|_\lambda$ converges to $\varepsilon'_1 = \beta'\varepsilon'_0/[1 - \beta'\bar{\rho}']$ where $\varepsilon'_0 = b_C b_B b_{Q_2}(b_C b_{x_0} + b_v)$. Based on (4.45), referring to the case of $t = 1$, the proof of Corollary 4.3.2 is thus finished by enumeration.

Remark 4.3.6. The effect of error. For simplicity, we discuss a special case when CB is square and with full rank. Instead of using an ideal $e_{i+1}(t+1)$ in (4.4), we use its prediction $\bar{e}_{i+1}(t+1)$ where

$$\bar{e}_{i+1}(t+1) = (I + \Gamma)e_{i+1}(t+1).$$

When $\Gamma = 0$, there is no prediction error. The prediction error can be absorbed into Q_0. To see a clear restriction on the prediction error Γ, we set

$$Q_0 = (I + \Gamma)Q_0^d(CB)^{-1},$$

where $Q_0^d = \text{diag}\{q_1, \cdots, q_m\}, \Gamma = \text{diag}\{\gamma_1, \cdots, \gamma_m\}$. Thus, from the definition of η and convergence condition (4.11), we need

$$\eta = \|[I + (I + \Gamma)Q_0^d(CB)^{-1}(CB)]^{-1}\|$$
$$\rightarrow |1 + (1 + \gamma_j)q_j|^{-1} < 1, \quad j = 1, \cdots, m.$$

So, for $q_j > 0$, we have

$$\gamma_j > -1, \text{ or, } \gamma_j < -2/q_j - 1.$$

Clearly, in practical systems, it is acceptable that when

$$|\gamma_j| < 1,$$

the proposed ILC scheme using CITE still works. This means that when the prediction error is within -100% to $+100\%$, the proposed scheme still works.

4.4 Simulation Illustrations

To demonstrate the effectiveness of the proposed CITE-assisted ILC algorithm, a single link direct joint driven manipulator model is used for the simulation study. The dynamic equation in the continuous-time t' domain is

$$\ddot{\theta}(t') = \frac{1}{J}[\tau(t') + \tau_n(t')] + \frac{1}{J}(\frac{1}{2}m_0 + M_0)gl\sin\theta(t') \tag{4.47}$$

where $\theta(t')$ is the angular position of the manipulator; $\tau(t')$ is the applied joint torque; $\tau_n(t')$ is the exogenous disturbance torque; m_0, l are the mass and length of the manipulator respectively, M_0 is the mass of the tip load, g is the gravitational acceleration and the J is the moment of inertia w.r.t the joint, i.e., $J = M_0l^2 + m_0l^2/3$. The parameters used in this simulation study are the same as those listed in Table 2.2.

Let the sampling period $T_s = 0.01$ sec. One can discretize the above model by using simple Euler method as follows:

$$\begin{cases} x_1(t+1) = x_2(t) \\ x_2(t+1) = 2x_2(t) - x_1(t) + [\tau(t) + \tau_n(t) + \\ \quad (0.5m_0 + M_0)gl\sin(x_1(t))]T_s^2/J \end{cases} \tag{4.48}$$

where discrete time $t = 0, 1, \cdots, 100$, $x_1(t) = \theta(t), x_2(t) = \theta(t+1)$. The desired tracking trajectories of the tracking tasks over the time interval $[0, 1]$ sec. are specified as

$$\theta_d(t) = \theta_b + (\theta_b - \theta_f)(15\hat{\tau}_0^4 - 6\hat{\tau}_0^5 - 10\hat{\tau}_0^3) \tag{4.49}$$

where $\hat{\tau}_0 = tT_s/(t_f - t_0)$. In the simulation, we use $\theta_b = 0°$, $\theta_f = 90°$, $t_0 = 0$, $t_f = 1$. The initial states at each ILC repetition are all set to 0. The ILC ends when $e_{b1} \leq 1°$ where $e_{b1} \overset{\triangle}{=} \sup_{t \in [0,100]} |\theta_d(t) - \theta(t)|$.

To apply the results obtained in this chapter, we examine the one-step forward shift scheme which makes $C(\cdot)B(\cdot) = 1$ as follows:

$$\tau_{i+1}(t) = \tau_i(t) + Q_0\bar{e}_{i+1}(t+2) + Q_1(e_i(t+2) - e_i(t+1)) \tag{4.50}$$

where $\bar{e}_{i+1}(t+1)$ is the prediction of $e_{i+1}(t+2)$ which is given by static extrapolation as

$$\begin{aligned} e_{i+1}(t+2) \approx \bar{e}_{i+1}(t+2) &\overset{\triangle}{=} y_d(t+1) - [3x_{2_{i+1}}(t-1) - \\ 2x_{2_{i+1}}(t-2)] &= y_d(t+1) - [3x_{1_{i+1}}(t) - 2x_{1_{i+1}}(t-1)]; \end{aligned} \tag{4.51}$$

Q_0, Q_1 is the learning parameters and $e(t) \overset{\triangle}{=} \theta_d(t) - \theta(t)$. In the ILC method, the system parameters are assumed unknown. However, the learning parameter determination should be based on the knowledge of the system. Fortunately, based on the the analysis of this chapter, this ILC learning parameter design difficulty can be relaxed. In the following simulation runs, set $Q_1 = 200$. We compared the ILC convergence processes in Fig. 4.1 for $Q_0 = 0, 25, 50, 75$. We can observe that even when the simplest 2-step extrapolation scheme is utilized, the CITE's effectiveness can be clearly illustrated by Fig. 4.1.

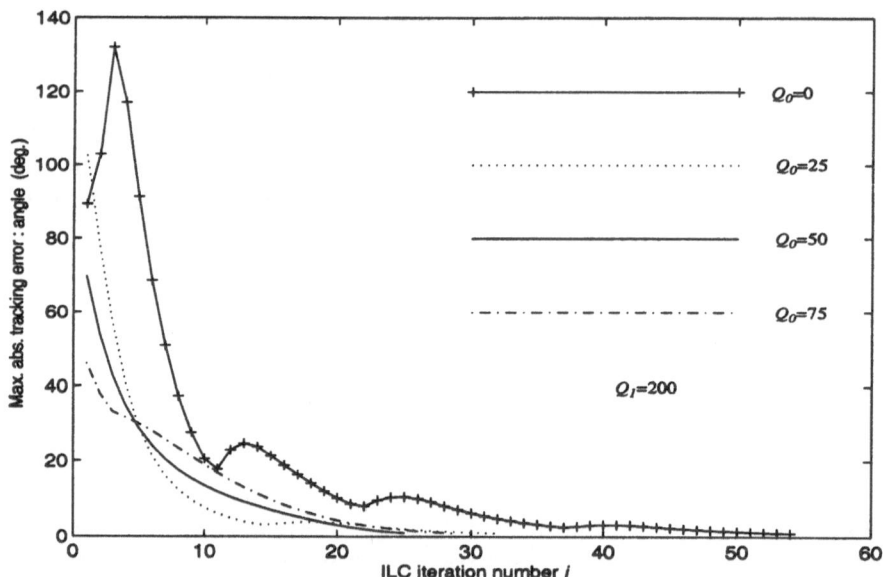

Fig. 4.1. Comparison of ILC convergence histories, ideal case with different CITE gains

It is more desirable to apply the ILC updating law with less forward time-shifts for the CITE. Instead of using CITE at $(t+2)$ instant (4.50), a $(t+1)$ version is as follows:

$$\tau_{i+1}(t) = \tau_i(t) + Q_0 \bar{e}_{i+1}(t+1) + Q_1(e_i(t+2) - e_i(t+1)) \qquad (4.52)$$

where $\bar{e}_{i+1}(t+1)$ is the prediction of $e_{i+1}(t+1)$ which is given by

$$
\begin{aligned}
e_{i+1}(t+1) &\approx \bar{e}_{i+1}(t+1) \overset{\triangle}{=} y_d(t+1) - [2x_{2_{i+1}}(t-1) - \\
x_{2_{i+1}}(t-2)] &= y_d(t+1) - [2x_{1_{i+1}}(t) - x_{1_{i+1}}(t-1)].
\end{aligned}
\qquad (4.53)
$$

More practically, one may prefer to use the ILC updating law

$$\tau_{i+1}(t) = \tau_i(t) + Q_0 \bar{e}_{i+1}(t-1) + Q_1(e_i(t+2) - e_i(t+1)) \qquad (4.54)$$

which is in the $(t-1)$ form for CITE usage. It is interesting to compare the three cases when ILC updating laws (4.50), (4.52), and (4.54) are applied respectively for a given Q_0, Q_1, say, $Q_0 = 25, Q_1 = 200$. The comparison is summarized in Fig. 4.2, which illustrates that the forward shifting of CITE is not crucial in this simulation. However, as shown in Fig. 4.2, the ILC convergence performance of (4.50) is the best. Fortunately, even when (4.54) is applied, the ILC convergence performance does not degrade significantly.

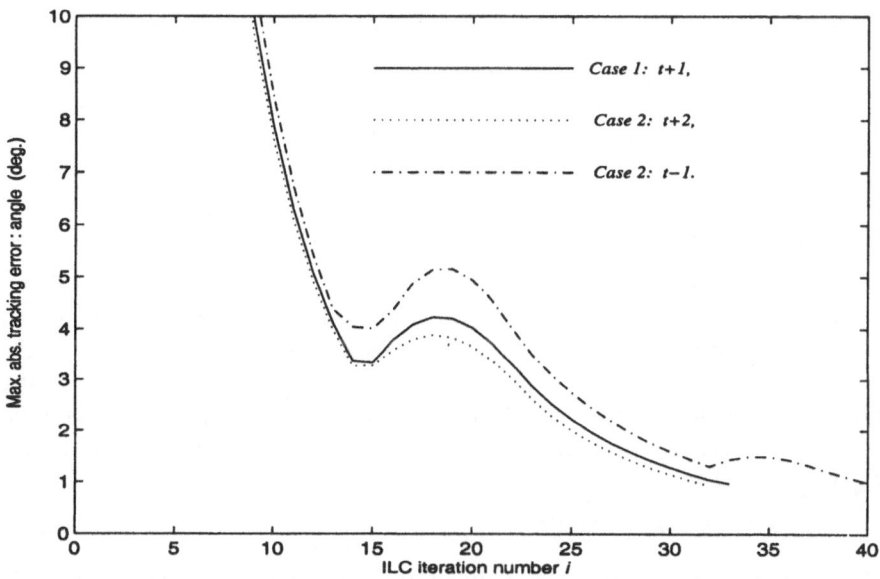

Fig. 4.2. Comparison of ILC convergence histories, ideal case with different CITE schemes (*Case 1: (4.50); Case 2: (4.52); Case 3: (4.54)*) (with local zoom)

As shown in Remark 4.3.6, when the sampling time is small and a suitable prediction scheme is applied, the proposed ILC scheme with CITE will work properly. This has been demonstrated in Fig. 4.2. The prediction errors for $t+1$ scheme (4.52) under $T_s = 0.01$s and a set of typical learning gains ($M = 1; Q_1 = 200, Q_0 = 25$) are shown in Fig. 4.3 for every ILC iteration. Similarly, for $t+2$ scheme (4.50), the errors are shown in Fig. 4.4. For different sampling periods the effects of prediction errors are summarized in Figs. 4.5-4.6 for scheme (4.52) and (4.50) respectively. From these calculations, it is quite clear that, in this simulation study, the prediction errors are not critical to the ILC convergence. This observation will also apply to many other applications.

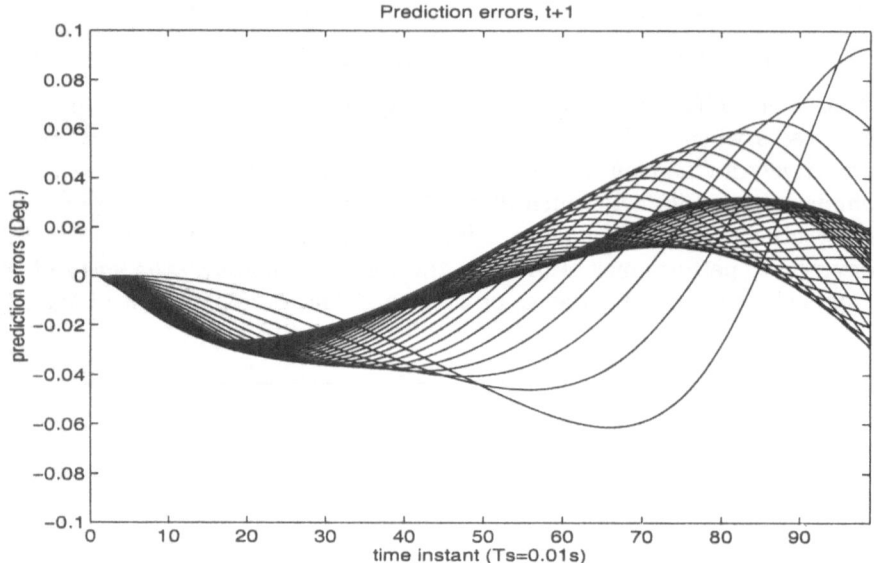

Fig. 4.3. Prediction errors in ILC. (t+1 version) ($T_s = 0.01s, Q_1 = 200, Q_0 = 25$).

Now we apply the CITE scheme (4.54) with $Q_1 = 200, Q_0 = 100$ for the tracking of the varying desired trajectories with exogenous torque disturbance. The desired trajectory is supposed to vary with respect to ILC iteration number i as discussed in Remarks 4.3.4 and 4.3.5. The varying desired trajectories are designed with the same form (4.49) but the final angle $\theta_f(t)$ is revised to

$$\theta_{f_i}(t) = (2 - e^{-0.05i})45°. \tag{4.55}$$

The varying desired trajectory is denoted by θ_{d_i}. The exogenous torque disturbance is supposed to be

$$\tau_n(t) = 30(2 - e^{-0.05i})\sin(\pi t/5) \text{ Nm.} \tag{4.56}$$

In the first ILC iteration, the desired trajectory $\theta_{d_0}(t)$ and the system output $\theta_0(t)$ are plotted in Fig. 4.7. To have a clear comparison, the desired trajectory and the system output at the 100-th ILC iteration, i.e., $\theta_{d_{100}}(t)$ and $\theta_{100}(t)$, are drawn in the same figure. We observe a good tracking in this case. It should be noted that in this case b_{y_d} defined in Remark 4.3.4 is fairly large. However, according to Remark 4.3.5, the tracking error bound in this

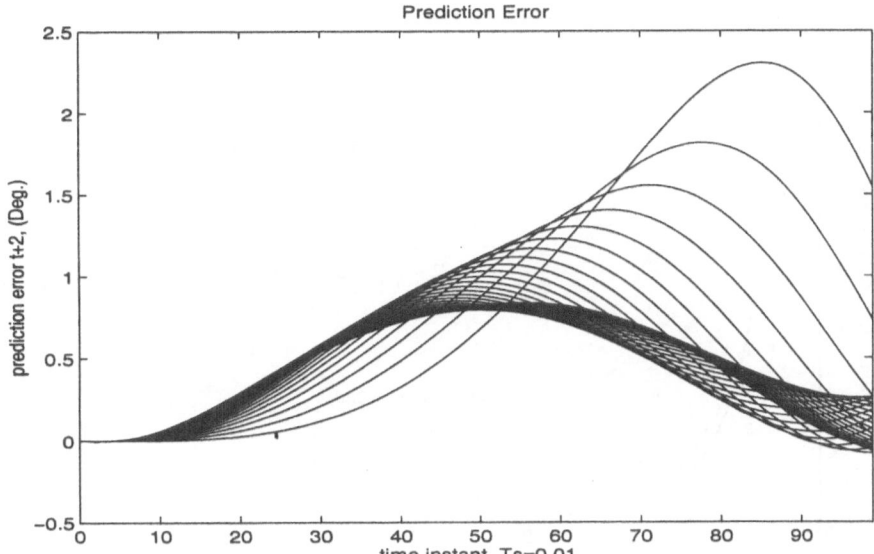

Fig. 4.4. Prediction errors in ILC. (t+2 version) ($T_s = 0.01s, Q_1 = 200, Q_0 = 25$).

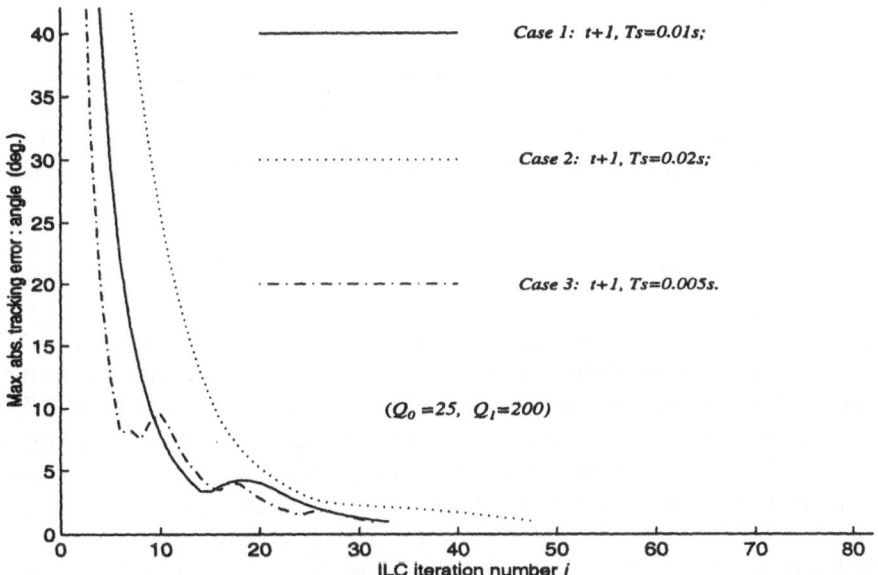

Fig. 4.5. Comparison of ILC convergence histories for different sampling periods. (t+1 version) (with local zoom) ($Q_1 = 200, Q_0 = 25$).

case should tend to 0 because b'_{y_d} tends to 0 as i increases. This is clearly illustrated by Fig. 4.7 which indicates that Remark 4.3.4 is more conservative than Remark 4.3.5. It should also be noted that in this case b_w defined

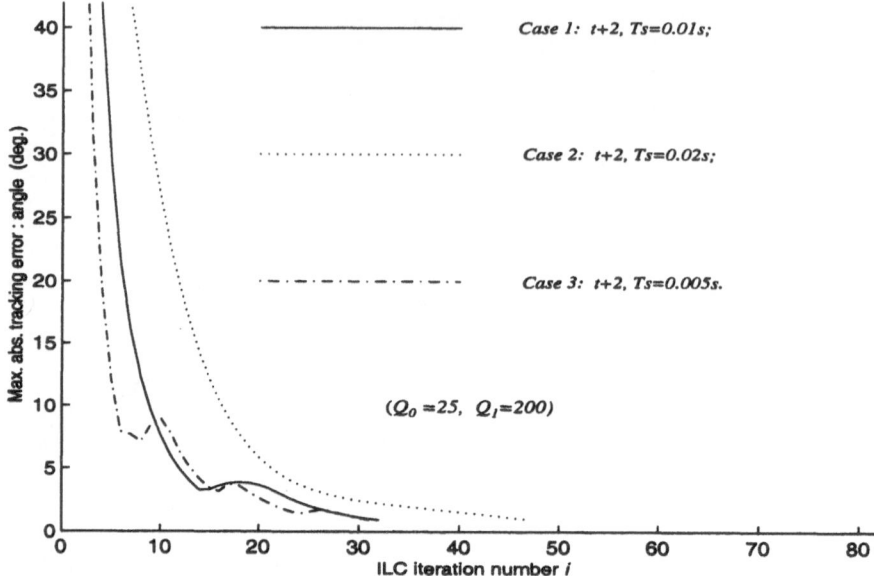

Fig. 4.6. Comparison of ILC convergence histories for different sampling periods. (t+2 version) (with local zoom) ($Q_1 = 200, Q_0 = 25$).

in Theorem 4.3.1 is fairly large. However, according to Theorem 4.3.2, the tracking error bound in this case should tend to 0 because b'_w tends to 0 as i increases. This is also clearly illustrated by Fig. 4.7 which indicates that the repetitive unknown exogenous disturbance can be rejected by ILC scheme.

4.5 Conclusion

An iterative learning controller, assisted by including a current iteration tracking error in its updating law, is proposed for uncertain discrete-time non-linear systems. It has been proven that the tracking error will be bounded in the presence of uncertainty, disturbance, and initialization error. Moreover, it has been shown that the final tracking error bound, which is directly related to the bounds of uncertainty, disturbance, and initialization error, together with the ILC convergence rate, can be adjusted to a desired level by tuning the learning gain of the current iteration tracking error in the ILC updating law. Simulation results have illustrated the effectiveness of the proposed iterative learning controller for tracking control of uncertain discrete-time nonlinear systems.

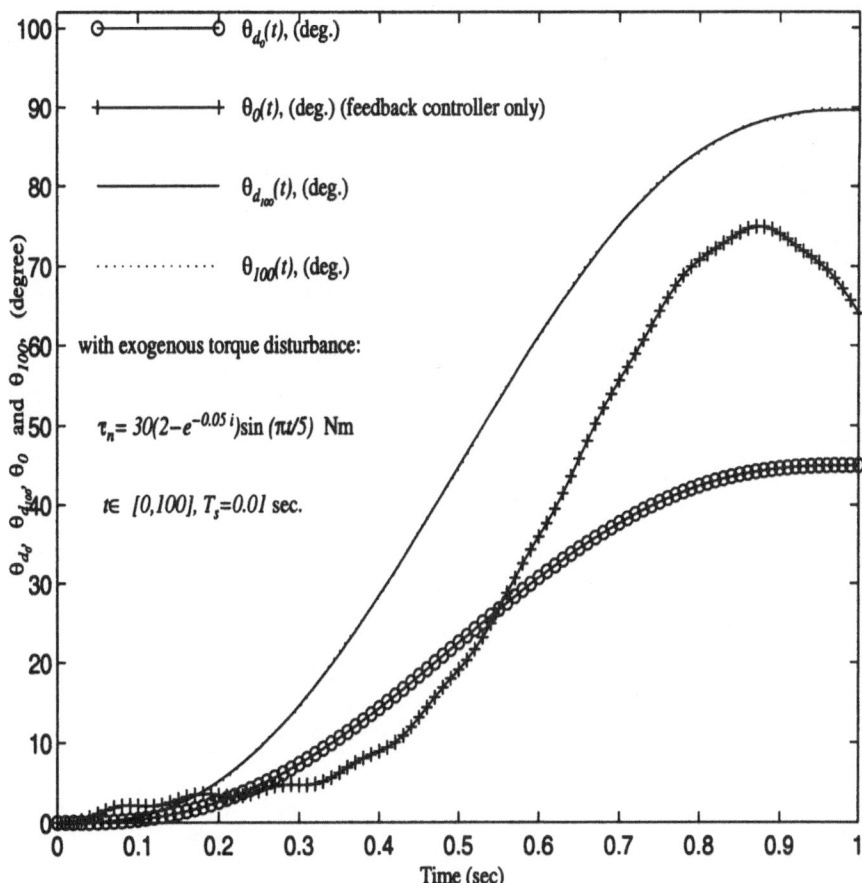

Fig. 4.7. Tracking the varying desired trajectory: with torque disturbance and CITE scheme *(4.54)*

5. Iterative Learning Control for Uncertain Nonlinear Discrete-time Feedback Systems With Saturation

5.1 Introduction

THE CONTROL FUNCTION from the ILC algorithm is applied to a system as an open-loop control and is updated by ILC updating law in a closed-loop fashion. From the robust analysis in [16], it is clear that if the trajectories of each iteration can be adjusted inside the neighborhood of the desired ones, the ILC performance will be better. This can be achieved by introducing a feedback loop. Then the system considered is actually controlled by an ILC feedforward controller and a feedback controller simultaneously. A block-diagram of the system is shown in Fig. 5.1. The use of a feedback controller plus the high-order scheme is mainly for the ILC performance improvements both in the time t-direction and in the ILC iteration number i-direction.

For the ILC scheme with a feedback controller, convergence results for continuous-time nonlinear systems using a first-order ILC can be found in [132, 120], but the robustness issues were not explored in detail. In the discrete-time case, there is still no convergence and robustness result reported for ILC schemes combined with a feedback controller. This chapter considers such problems. In the analysis, a high-order ILC updating law plus a feedback controller is used for tracking control of uncertain discrete-time nonlinear systems. The effect of control input saturation is also addressed. The feedback controller considered in this chapter is in a general form including the scheme addressed in Chapter 4 as a special case. Uniform boundedness of the tracking error is established in the presence of bounded uncertainty, disturbance and re-initialization error even without the assistance of a feedback controller. It is also shown that the tracking error bound is a function of the bounds of the differences of uncertainties, disturbances and the re-initialization errors between two successive ILC iterations under some additional conditions.

The organization of this chapter is as follows. In Sec. 5.2, the control problem is formulated with the proposed 'high-order ILC plus feedback controller' scheme. The main results of the ILC convergence and robustness are given in Sec. 5.3 for bounded uncertainty and disturbance. Extension to a more general ILC updating law is also discussed in Sec. 5.3. In Sec. 5.3.2, the robustness against the difference bounds of disturbance or uncertainty between two successive ILC iterations is addressed. Simulation illustrations are presented in Sec. 5.4 to verify the effectiveness of the proposed schemes. Finally, chapter summaries are drawn in Sec. 5.5.

5.2 Problem Settings

5.2.1 Discrete Uncertain Time-varying Nonlinear Systems

Consider the following discrete-time uncertain nonlinear time-varying system which performs a given task repeatedly.

$$\begin{cases} x_i(t+1) = f(x_i(t), t) + B(x_i(t), t)u_i(t) + w_i(t) \\ y_i(t) = C(t)x_i(t) + v_i(t) \end{cases} \tag{5.1}$$

where i denotes the i-th repetitive operation of the system; t is the discrete time index and $t \in [0, N]$ which means that $t \in \{0, 1, \cdots, N\}$; $x_i(t) \in R^n$, $u_i(t) \in R^m$, and $y_i(t) \in R^r$ are the state, control input, and output of the system, respectively; $C(t) \in R^{r \times n}$ is a time-varying matrix; the functions $f(\cdot, \cdot) : R^n \times [0, N] \mapsto R^n$ and $B(\cdot, \cdot) : R^n \times [0, N] \mapsto R^m$ are uniformly globally Lipschitzian in x, i.e.,$\forall t \in [0, N], \forall i, \exists$ constants k_f, k_B, such that

$$\|\Delta f_i(t)\| \le k_f \|\Delta x_i(t)\|, \quad \|\Delta B_i(t)\| \le k_B \|\Delta x_i(t)\|$$

where $\Delta f_i(t) \triangleq f(x_i(t), t) - f(x_{i-1}(t), t)$, $\Delta B_i(t) \triangleq B(x_i(t), t) - B(x_{i-1}(t), t)$, $\Delta x_i(t) \triangleq x_i(t) - x_{i-1}(t)$; $w_i(t), v_i(t)$ are uncertainty or disturbance to the system bounded with unknown bounds b_w, b_v defined as

$$b_w \triangleq \sup_{t \in [0, N]} \|w_i(t)\|, \quad b_v \triangleq \sup_{t \in [0, N]} \|v_i(t)\|, \ \forall i. \tag{5.2}$$

Denote the output tracking error $e_i(t) \triangleq y_d(t) - y_i(t)$ where $y_d(t)$ is the given desired output trajectory, which is realizable, i.e., given a bounded $y_d(t)$, there exists a unique bounded desired input $u_d(t)$, $t \in [0, N]$ such that when $u(t) = u_d(t)$, the system has a unique bounded desired state $x_d(t)$ satisfying

$$\begin{cases} x_d(t+1) = f(x_d(t), t) + B(x_d(t), t)u_d(t) \triangleq f_d + B_d u_d \\ y_d(t) = C(t)x_d(t) \triangleq C(t)x_d. \end{cases} \tag{5.3}$$

Denote the bound of the desired control u_d as $b_{u_d} \triangleq \sup_{t \in [0, N]} \|u_d(t)\|$.

5.2.2 Discrete-time High-order ILC with a Feedback Controller

The control problem is formulated as follows. Starting from an arbitrary initial control input $u_0(t)$, obtain the next control input $u_1(t)$ and the subsequent series $\{u_i(t) \mid i = 2, 3, \cdots\}$ for system (5.1) by using a proper learning control updating law in such a way that when $i \to \infty, y_i(t) \to y_d(t) \pm \epsilon^*$ in the presence of bounded uncertainty, disturbance and re-initialization error.

To solve the above problem, we propose a high-order iterative learning controller combined with a feedback controller in general form as illustrated in Fig. 5.1.

At the i-th ILC iteration, suppose that the control input $u_i(t)$ to the system (5.1) is the output of a saturater, i.e.,

$$u_i(t) = \text{sat}(\bar{u}_i(t)) \tag{5.4}$$

where $\text{sat}(\bar{u}_i(t)) \triangleq [\text{sat}(\bar{u}_{1_i}(t)), \cdots, \text{sat}(\bar{u}_{m_i}(t))]^T$ and

$$\text{sat}(\bar{u}_{j_i}(t)) \triangleq \begin{cases} \bar{u}_{j_i}(t), & \text{if } | \bar{u}_{j_i}(t) | \leq \bar{u}_j^* \\ \frac{\bar{u}_{j_i}(t)}{|\bar{u}_{j_i}(t)|} \bar{u}_j^*, & \text{if } | \bar{u}_{j_i}(t) | > \bar{u}_j^* \end{cases} \tag{5.5}$$

where $j = 1, 2, \cdots, m$ and $\bar{u}_j^* > 0$ are the saturation bounds. The introduction of an input saturater here is to show that the ILC is also effective in such realistic cases as argued in [120]. The saturater input is that

$$\bar{u}_i(t) = u_i^f(t) + u_i^b(t) \tag{5.6}$$

where $u_i^f(t)$ is from the feedforward iterative learning controller and $u_i^b(t)$ is from the feedback stabilizing controller. The feedback stabilizing controller is assumed to be in the following general form.

$$z_i(t+1) = h_a(z_i(t)) + H_b(z_i(t))e_i(t), \tag{5.7}$$

$$u_i^b(t) = h_c(z_i(t)) + H_d(z_i(t))e_i(t) \tag{5.8}$$

where $z_i(t) \in R^{n_c}$ is the state of the feedback stabilizing controller with $z_i(0) = 0, \forall i$. The vector-valued functions $h_a(\cdot) : R^{n_c} \mapsto R^{n_c}$ and $h_c(\cdot) : R^{n_c} \mapsto R^m$ are designed to be sector-bounded as

$$\|h_a(z_i(t))\| \leq b_{h_a}\|z_i(t)\|, \ \|h_c(z_i(t))\| \leq b_{h_c}\|z_i(t)\|.$$

The function matrices $H_b(\cdot) : R^{n_c} \mapsto R^{n_c \times r}$ and $H_d(\cdot) : R^{n_c} \mapsto R^{m \times r}$ are designed to be with uniform bounds, i.e., $\forall t \in [0, N], \forall z_i(t) \in R^{n_c}$,

$$\|H_b(z_i(t))\| \leq b_{H_b}, \ \|H_d(z_i(t))\| \leq b_{H_d}.$$

The positive constants $b_{h_a}, b_{h_c}, b_{H_b}, b_{H_d}$ are not necessarily known. A simple ILC updating law is used which includes tracking errors of M previous iterations, i.e.,

$$u_{i+1}^f(t) = u_i(t) + \sum_{k=1}^{M} Q_k(t)e_{i-k+1}(t+1) \tag{5.9}$$

where M is the order of the ILC updating law with $M \geq 1$; $Q_k(t) \in R^{m \times r}(k = 1, \cdots, M)$ are the learning matrices which are to be determined to ensure the ILC convergence.

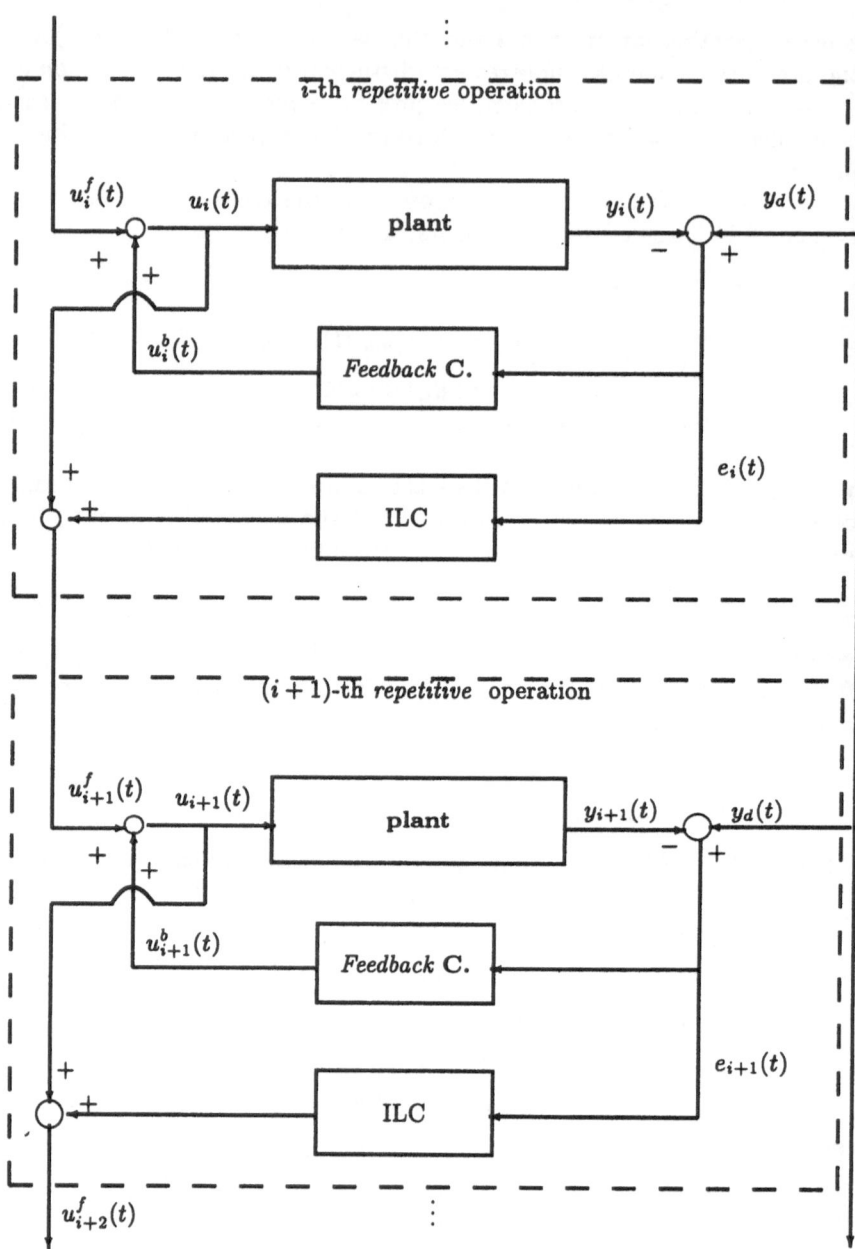

Fig. 5.1. Block-diagram of an iterative learning controller assisted by a feedback controller

Remark 5.2.1. Because the tracking errors of previous ILC iterations are pre-stored in the memory and can be manipulated easily to shift the time index by 1, the above proposed ILC updating law (5.9) can be taken as a PD-type scheme because $e_i(t+1) \approx e_i(t) + T_s \dot{e}_i(t)$ where T_s is the sampling time. Thus the scheme (5.9) is essentially PD-PID type.

Remark 5.2.2. The ILC scheme with CITE considered in Chapter 4 is a special case of the schemes of this chapter as illustrated by Fig. 5.1.

5.2.3 Assumptions and Preliminaries

To restrict our discussion, the following assumptions are made.

- A1). The initialization error is bounded as follows, $\forall t \in [0, N]$, $\forall i$, $\|x_d(0) - x_i(0)\| \le b_{x_0}$, $\|y_d(0) - y_i(0)\| \le b_C b_{x_0} + b_v$, where $b_C \stackrel{\triangle}{=} \sup_{t \in [0,N]} \|C(t)\|$.
- A2). Matrix $C(\cdot)B(\cdot, \cdot)$ has a full column rank $\forall t \in [0, N]$, $x(t) \in R^n$.
- A3). Operator $B(\cdot, \cdot)$ is bounded, i.e., $\exists$ a constant b_B such that for all i, $\sup_{t \in [0,N]} \|B(x_i(t), t)\| \stackrel{\triangle}{=} \sup_{t \in [0,N]} \|B_i(t)\| \le b_B$.
- A4). The desired output $y_d(t), t \in [0, N]$ is achievable by the desired input $u_d(t)$ and $u_d(t)$ is within the saturation bounds, i.e., $u_d(t) \equiv \text{sat}(u_d(t)), \forall t \in [0, N]$.

Assumption A1) restricts that the initial states or the initial outputs in each repetitive operation should be inside a given ball centered at the desired initial ones. The radius of the ball may be unknown. The number of outputs r can be greater than or equal to the number of inputs m according to A2). A3) indicates that the range of operator $B(\cdot, \cdot)$ is always finite. This is reasonable because the repetitive tasks are performed in a finite time interval $[0, NT_s]$. A4) requires that the desired trajectory should be planned in such a way that those large and sudden changes, which the system can not follow, should be excluded from the desired trajectories.

To facilitate the later derivations, some basic relations are presented in the following. The main purpose is to explore the relationship between $(\|\delta x_i(t)\|_\lambda + \|z_i(t)\|_\lambda)$ and $\|\delta u_i^f(t)\|_\lambda$ where $\delta u_i^f(t) \stackrel{\triangle}{=} u_d(t) - u_i^f(t)$, $\delta x_i(t) \stackrel{\triangle}{=} x_d(t) - x_i(t)$. Denote $\delta u_i(t) \stackrel{\triangle}{=} u_d(t) - u_i(t)$. Similar to the λ-norm defined in (4.5), the constant $\hat{e}$ will be specified later.

Denote

$$\delta f_i(t) \stackrel{\triangle}{=} f_d - f(x_i(t), t), \quad \delta B_i(t) \stackrel{\triangle}{=} B_d - B_i(t).$$

Then, from (5.1) and (5.3), it can be obtained that

$$\delta x_i(t+1) = \delta f_i(t) + \delta B_i(t)u_d + B_i(t)\delta u_i(t) - w_i(t). \tag{5.10}$$

Taking the norm for (5.10) yields

$$\|\delta x_i(t+1)\| \le (k_f + b_{u_d}k_B)\|\delta x_i(t)\| + b_B\|\delta u_i(t)\| + b_w. \tag{5.11}$$

From (5.8), it can be seen that

$$\|u_i^b(t)\| \le b_{h_c}\|z_i(t)\| + b_{H_d}b_C\|\delta x_i(t)\| + b_{H_d}b_v. \tag{5.12}$$

By noticing that $u_d(t) \equiv \text{sat}(u_d(t))$ and that

$$
\begin{aligned}
\|\delta u_i(t)\| &= \|u_d(t) - \text{sat}(u_i^f(t) + u_i^b(t))\| \\
&\leq \|\delta u_i^f(t) - u_i^b(t)\| \leq \|\delta u_i^f(t)\| + \|u_i^b(t)\|,
\end{aligned} \tag{5.13}
$$

then, (5.11) becomes

$$
\begin{aligned}
\|\delta x_i(t+1)\| &\leq (k_f + b_{u_d}k_B + b_B b_{H_d} b_C)\|\delta x_i(t)\| \\
&+ b_B b_{h_c}\|z_i(t)\| + b_B\|\delta u_i^f(t)\| + b_B b_{H_d} b_v + b_w.
\end{aligned} \tag{5.14}
$$

On the other hand, it can be observed from (5.7) that

$$
\|z_i(t+1)\| \leq b_{h_a}\|z_i(t)\| + b_{H_b} b_C\|\delta x_i(t)\| + b_{H_b} b_v. \tag{5.15}
$$

Thus, adding (5.15) into (5.14) yields

$$
\begin{aligned}
(\|\delta x_i(t+1)\| + \|z_i(t+1)\|) &\leq \hat{e}(\|\delta x_i(t)\| + \|z_i(t)\|) \\
&+ b_B\|\delta u_i^f(t)\| + \hat{\varepsilon}
\end{aligned} \tag{5.16}
$$

where

$$
\begin{aligned}
\hat{e} &\triangleq \max\{k_f + b_{u_d}k_B + b_B b_{H_d} b_C + b_{H_b} b_C, \ b_{h_a} + b_B b_{h_c}\} \neq 1; \\
\hat{\varepsilon} &\triangleq (b_{H_b} + b_B b_{H_d})b_v + b_w.
\end{aligned}
$$

Applying (4.6), we can get

$$
\begin{aligned}
\|\delta x_i(t+1)\| + \|z_i(t+1)\| &\leq \hat{e}^{t+1} b_{x_0} \\
&+ \sum_{j=0}^{t} \hat{e}^{t-j}(b_B\|\delta u_i^f(j)\| + \hat{\varepsilon}).
\end{aligned} \tag{5.17}
$$

To see a simpler relationship between $\|\delta x_i(t)\|_\lambda + \|z_i(t)\|_\lambda$ and $\|\delta u_i^f(t)\|_\lambda$, by noticing the properties of $\lambda-$norm presented in Chapter 4, then, taking the $\lambda-$norm $(|\lambda| > 1)$ operation of (5.17) gives

$$
\|\delta x_i(t)\|_\lambda + \|z_i(t)\|_\lambda \leq b_{x_0} + b_B O(|\lambda|^{-1})\|\delta u_i^f(t)\|_\lambda + c_0\hat{\varepsilon} \tag{5.18}
$$

where

$$
O(|\lambda|^{-1}) \triangleq \frac{1 - \hat{e}^{-(\lambda-1)N}}{\hat{e}^\lambda - \hat{e}}, \quad c_0 \triangleq \sup_{t \in [0,N]} \frac{\hat{e}^{-(\lambda-1)t}(1 - \hat{e}^{-t})}{\hat{e} - 1}.
$$

For brevity of our discussion, in the sequel, the following notations are used.

$$
b_{Q_k} \triangleq \sup_{t \in [0,N]} \|Q_k(t)\|, \quad (k = 1, 2, \cdots, M),
$$

$$
\rho_1 \triangleq \sup_{t \in [0,N]} \|I_m - Q_1(t)C(t+1)B_i(t)\|, \quad \forall i,
$$

$$
\rho_k \triangleq \sup_{t \in [0,N]} \|Q_k(t)C(t+1)B_{i-k+1}(t)\|, \quad \forall i, \ (k = 2, 3, \cdots, M).
$$

5.3 Robust Convergence Analysis

5.3.1 Robustness Against Bounded Uncertainty and Disturbance

A main result on the robustness against bounded uncertainty and disturbance
is presented in the following theorem.

Theorem 5.3.1 *For the repetitive discrete-time uncertain time-varying nonlinear system (5.1) under assumptions A1)-A4), given the realizable desired
trajectory $y_d(t)$ over the fixed time interval $[0, NT_s]$, by using the ILC updating law (5.9) and the feedback controller (5.7)-(5.8), if the condition*

$$\rho \triangleq \sum_{k=1}^{M} \rho_k < 1, \tag{5.19}$$

*is satisfied, then the λ-norm of the tracking errors $e_i(t)$, $\delta u_i(t)$, $\delta x_i(t)$ are
bounded for all i. For a sufficiently large $|\lambda|$, $\forall t \in [0, N]$,*

$$b_{u^f} \triangleq \lim_{i \to \infty} \|\delta u_i^f(t)\|_\lambda \le b_{u^f}(b_{x_0}, b_w, b_v), \tag{5.20}$$

$$b_u \triangleq \lim_{i \to \infty} \|\delta u_i(t)\|_\lambda \le b_u(b_{x_0}, b_w, b_v), \tag{5.21}$$

$$b_x \triangleq \lim_{i \to \infty} \|\delta x_i(t)\|_\lambda \le b_{x_0} + b_B O(|\lambda|^{-1}) b_{u^f} + c_0 \hat{\varepsilon}, \tag{5.22}$$

$$b_e \triangleq \lim_{i \to \infty} \|e_i(t)\|_\lambda \le b_C b_x + b_v. \tag{5.23}$$

*Moreover, b_u, b_x, b_e will all converge uniformly to zero for $t = 0, 1, \cdots, N$ as
$i \to \infty$ in the absence of uncertainty, disturbance and initialization error,
i.e., $b_w, b_v, b_{x_0} \to 0$.*

Proof. The tracking error at $(i + 1)$−th repetition is

$$\begin{aligned} e_i(t) &= y_d(t) - y_i(t) \\ &= C(t)\delta x_i(t) - v_i(t). \end{aligned} \tag{5.24}$$

Investigating the learning control deviation at the $(i + 1)$-th repetition
$\delta u_{i+1}^f(t)$ gives

$$\delta u_{i+1}^f(t) = \delta u_i(t) - \sum_{k=1}^{M} Q_k(t) e_{i-k+1}(t+1)$$

$$= \delta u_i(t) - \sum_{k=1}^{M} Q_k(t) C(t+1) \delta x_{i-k+1}(t+1)$$

$$+ \sum_{k=1}^{M} Q_k(t) v_{i-k+1}(t+1). \tag{5.25}$$

By referring to (5.10), (5.25) can be written as

$$\delta u_{i+1}^f(t) = \delta u_i(t) - \sum_{k=1}^{M} Q_k(t) C(t+1)$$

$$[\delta f_{i-k+1}(t) + \delta B_{i-k+1}(t) u_d + B_{i-k+1}(t) \delta u_{i-k+1}(t)$$

$$-w_{i-k+1}(t)] + \sum_{k=1}^{M} Q_k(t) v_{i-k+1}(t+1). \tag{5.26}$$

Collecting terms and then performing the norm operation for (5.26) yield

$$\|\delta u_{i+1}^f(t)\| \le \sum_{k=1}^{M} \rho_k \|\delta u_{i-k+1}(t)\|$$

$$+ \sum_{k=1}^{M} b_{Q_k} b_C (k_f + b_{u_d} k_B) \|\delta x_{i-k+1}(t)\|$$

$$+ \sum_{k=1}^{M} b_{Q_k} (b_C b_w + b_v). \tag{5.27}$$

Based on (5.13) and (5.12), (5.27) becomes

$$\|\delta u_{i+1}^f(t)\| \le \sum_{k=1}^{M} \rho_k \|\delta u_{i-k+1}^f(t)\|$$

$$+ \sum_{k=1}^{M} \alpha_k (\|\delta x_{i-k+1}(t)\| + \|z_{i-k+1}(t)\|) + \varepsilon. \tag{5.28}$$

where

$$\alpha_k \triangleq \max\{b_{Q_k} b_C (k_f + b_{u_d} k_B) + b_{H_d} b_C \rho_k, b_{h_c} \rho_k\}$$

and

$$\varepsilon \triangleq \sum_{k=1}^{M} [b_{Q_k} (b_C b_w + b_v) + b_{H_d} b_v \rho_k].$$

By utilizing the relationship in (5.18), taking the $\lambda-$norm for (5.27) gives

$$\|\delta u_{i+1}^f(t)\|_\lambda \le \sum_{k=1}^{M} \rho_k \|\delta u_{i-k+1}^f(t)\|_\lambda$$

$$+ \sum_{k=1}^{M} \alpha_k b_B O(|\lambda|^{-1}) \|\delta u_{i-k+1}^f(t)\|_\lambda$$

$$+ \sum_{k=1}^{M} \alpha_k (b_{x_0} + c_0 \hat{\varepsilon}) + \varepsilon. \tag{5.29}$$

Referring to (5.19), it is clear that a sufficiently large $|\lambda|$ can be used to ensure that

$$\sum_{k=1}^{M}(\rho_k + \alpha_k b_B O(|\lambda|^{-1})) \triangleq \hat{\rho} < 1, \tag{5.30}$$

therefore, according to Lemma 2.3.1, we have

$$b_{u^f} = \lim_{i\to\infty} \|\delta u_i^f(t)\|_\lambda = \frac{\varepsilon_0}{1-\hat{\rho}} \triangleq b_{u^f}(b_{x_0}, b_w, b_v) \tag{5.31}$$

where $\varepsilon_0 \triangleq \varepsilon + \sum_{k=1}^{M}\alpha_k(b_{x_0}+c_0\hat{\varepsilon})$. From (5.18) and (5.24), (5.22) and (5.23) can be verified. It can be observed from (5.18) that

$$b_{xz} \triangleq \lim_{i\to\infty}(\|\delta x_i(t)\|_\lambda + \|z_i(t)\|_\lambda)$$
$$\leq b_{x_0} + b_B O(|\lambda|^{-1})b_{u^f} + c_0\hat{\varepsilon}. \tag{5.32}$$

Therefore, from (5.13) and (5.12), by referring to (5.32), we have

$$b_u \triangleq \lim_{i\to\infty}\|\delta u_i(t)\|_\lambda \leq b_{u^f} + \max\{b_{h_c}, b_{H_d}b_C\}b_{xz} + b_{H_d}b_v$$
$$\triangleq b_u(b_{x_0}, b_w, b_v) \tag{5.33}$$

which verifies (5.21). Moreover, it is easy to observe that b_{u^f}, b_u, b_x, and b_e will all tend to zero uniformly for $t = 0, 1, \cdots, N$ as $i \to \infty$ in the absence of uncertainty, disturbance and initialization error, i.e., when $b_w, b_v, b_{x_0} \to 0$.

Remark 5.3.1. Consider the case that the desired trajectory varies with respect to ILC iteration number i. Suppose the desired trajectory at the i-th iteration is changed to $y_{d_i}(t)$. If

$$\|y_{d_i}(t) - y_d(t)\| < b_{y_d}, \quad \forall t \in [0, N] \text{ and } \forall i, \tag{5.34}$$

then, all the discussions made above are all valid by replacing b_v with $b_v + b_{y_d}$.

The tracking errors in the ILC updating law (5.9) are shifted by one sampling step, i.e., $e_{i-k+1}(t+1)$. In the following Corollary, we will show that the result of Theorem 5.3.1 still holds if tracking errors $e_{i-k+1}(t)$ are included in (5.9), i.e., when the ILC updating law

$$u_{i+1}^f(t) = u_i(t) + \sum_{k=1}^{M}[Q_k(t)e_{i-k+1}(t+1) + \bar{Q}_k(t)e_{i-k+1}(t)] \tag{5.35}$$

is applied, where $\bar{Q}_k(t)$ are the bounded learning matrices $\forall t \in [0, N]$.

Corollary 5.3.1 *The ILC convergence in Theorem 5.3.1 still holds when the ILC updating law (5.9) is replaced by the ILC updating law (5.35).*

Proof. The proof is quite similar to the proof of Theorem 5.3.1. When (5.35) is considered, the following term

$$-\sum_{k=1}^{M} \bar{Q}_k(t) e_{i-k+1}(t) + \sum_{k=1}^{M} \bar{Q}_k(t) v_{i-k+1}$$

should be appended in the left-hand sides of (5.25) and (5.26). Moreover, based on (5.24), we get the following inequality of the norm estimates similar to (5.27).

$$\|\delta u_{i+1}^f(t)\| \le \sum_{k=1}^{M} \rho_k \|\delta u_{i-k+1}(t)\|$$

$$+\sum_{k=1}^{M} [b_{Q_k}(k_f + b_{u_d} k_B) + b_{\bar{Q}_k} b_C] \|\delta x_{i-k+1}(t)\|$$

$$+\sum_{k=1}^{M} [b_{Q_k}(b_C b_w + b_v) + b_{\bar{Q}_k} b_v] \tag{5.36}$$

where $b_{\bar{Q}_k} \triangleq \sup_{t \in [0,N]} \|\bar{Q}_k(t)\|$. The remaining part of the proof is the same as the proof of Theorem 5.3.1 if we set

$$\alpha_k \triangleq \max\{b_{Q_k}(k_f + b_{u_d} k_B) + b_{\bar{Q}_k} b_C + b_{H_d} b_C \rho_k, b_{h_c} \rho_k\}$$

and

$$\varepsilon \triangleq \sum_{k=1}^{M} [b_{Q_k}(b_C b_w + b_v) + b_{\bar{Q}_k} b_v + b_{H_d} b_v \rho_k],$$

which ends the proof.

It can be observed that the convergence of the discrete-time ILC is only governed by the learning components $Q_k(t) e_{i-k+1}(t+1)$. However, the additional terms $\bar{Q}_k(t) e_{i-k+1}(t)$ in the ILC updating law may improve the performance of ILC convergence as argued in the following Remark.

Remark 5.3.2. Refer to Remark 5.2.1, if we set $\bar{Q}_k(t) = -Q_k(t)$, it is obvious that the result of Corollary 5.3.1 is a discrete analogy of the result of [120] when the relative degrees of the nonlinear system are all one.

We now extend the above obtained results for a more general ILC updating law. In the ILC updating law (5.9), only the tracking errors of previous cycles were utilized. The control functions of previous cycles can also be utilized also as in the high-order ILC schemes [27, 46]. The general form of the high-order ILC updating law is given as

$$u_{i+1}^f(t) = \sum_{k=1}^{M} [P_k(t) u_{i-k+1}(t) + Q_k(t) e_{i-k+1}(t+1)$$

$$+\bar{Q}_k(t) e_{i-k+1}(t)] \tag{5.37}$$

where the learning matrices $P_k(t)$ $(k = 1, 2, \cdots, M)$ satisfy

$$\sum_{k=1}^{M} P_k(t) = I_m, \quad \forall t \in [0, N]. \tag{5.38}$$

We can get a similar robust convergence result to Theorem 5.3.1.

Theorem 5.3.2 *Consider the repetitive discrete-time uncertain time-varying nonlinear system (5.1) satisfying Assumptions A1)-A4). For a given realizable desired trajectory $y_d(t)$ over the fixed time interval $[0, NT_s]$, the iterative learning controller (5.37) and feedback controller (5.7)-(5.8) are applied. If*

$$\rho' \triangleq \sum_{k=1}^{M} \rho_k' < 1, \tag{5.39}$$

where $\forall i$, and $k = 1, 2, \cdots, M$,

$$\rho_k' \triangleq \sup_{t \in [0,N]} \| P_k(t) - Q_k(t)C(t+1)B_{i-k+1}(t) \|, \tag{5.40}$$

then the λ-norm of the tracking errors $e_i(t)$, $\delta u_i^f(t)$, $\delta u_i(t)$, $\delta x_i(t)$ are all bounded for all i. The bounds of the tracking errors are functions of b_{x_0}, b_w, b_v and moreover, will all converge uniformly to zero for $t = 0, 1, \cdots, N$ as $i \to \infty$ in the absence of uncertainty, disturbance and initialization error, i.e., $b_w, b_v, b_{x_0} \to 0$.

Proof. The main body of the proof is similar to the proof of Theorem 5.3.1 where (5.25) and (5.26) should be re-written as

$$\delta u_{i+1}^f(t) = u_d(t) - \sum_{k=1}^{M} P_k(t)u_{i-k+1}(t)$$

$$- \sum_{k=1}^{M} [Q_k(t)e_{i-k+1}(t+1) + \bar{Q}_k(t)e_{i-k+1}(t)]$$

$$= u_d(t) - \sum_{k=1}^{M} P_k(t)u_{i-k+1}(t)$$

$$- \sum_{k=1}^{M} Q_k(t)C(t+1)\delta x_{i-k+1}(t+1)$$

$$+ \sum_{k=1}^{M} Q_k(t)v_{i-k+1}(t+1)$$

$$- \sum_{k=1}^{M} \bar{Q}_k(t)C(t)\delta x_{i-k+1}(t) + \sum_{k=1}^{M} \bar{Q}_k(t)v_{i-k+1}(t). \tag{5.41}$$

By referring to (5.10) and (5.38), (5.41) can be written as

$$\delta u_{i+1}^{f}(t) = \sum_{k=1}^{M} P_k(t)\delta u_{i-k+1}(t) - \sum_{k=1}^{M} Q_k(t)C(t+1)$$

$$[\delta f_{i-k+1}(t) + \delta B_{i-k+1}(t)u_d + B_{i-k+1}(t)\delta u_{i-k+1}(t)$$

$$-w_{i-k+1}(t)] + \sum_{k=1}^{M} Q_k(t)v_{i-k+1}(t+1)$$

$$-\sum_{k=1}^{M} \bar{Q}_k(t)C(t)\delta x_{i-k+1}(t) + \sum_{k=1}^{M} \bar{Q}_k(t)v_{i-k+1}(t). \qquad (5.42)$$

Estimating the norm yields

$$\|\delta u_{i+1}^{f}(t)\| \le \sum_{k=1}^{M} \rho_k' \|\delta u_{i-k+1}(t)\|$$

$$+\sum_{k=1}^{M}[b_{Q_k}(k_f + b_{u_d}k_B) + b_{\bar{Q}_k}b_C]\|\delta x_{i-k+1}(t)\|$$

$$+\sum_{k=1}^{M}[b_{Q_k}(b_C b_w + b_v) + b_{\bar{Q}_k}b_v]. \qquad (5.43)$$

The remaining part of the proof is the same as the proof of Theorem 5.3.1 if we set

$$\alpha_k \overset{\triangle}{=} \max\{b_{Q_k}(k_f + b_{u_d}k_B) + b_{\bar{Q}_k}b_C + b_{H_d}b_C\rho_k', b_{h_c}\rho_k'\}$$

and

$$\varepsilon \overset{\triangle}{=} \sum_{k=1}^{M}[b_{Q_k}(b_C b_w + b_v) + b_{\bar{Q}_k}b_v + b_{H_d}b_v\rho_k'],$$

which ends the proof.

Remark 5.3.3. The general form of high-order ILC updating law (5.37) introduces more flexibility in determining the learning matrices. This can be taken as a form of "filtered input" case as considered in [144]. However, the increased number of design parameters will obscure the expected advantages. Thus, from practical point of view, only using dual PIDs both in the t-axis and in the i-direction for the proposed feedback-assisted ILC will be attractive based on the long-history usage of the PID controller in industries as discussed in Chapter 2.

5.3.2 Robustness Against Uncertainty or Disturbance Difference

In the following we will show that under certain additional restrictions, the tracking error bound is a function of the bounds of the differences of uncertainties, disturbances and the re-initialization errors between two successive ILC iterations even without a feedback controller, i.e., $u_i(t) \equiv u_i^{f}(t)$.

Introduce a similar $\lambda-$norm definition as in (4.5) with $\hat{e}$ replaced by

$$\tilde{e} \stackrel{\Delta}{=} k_f + b_{u^*} k_B \neq 1 \tag{5.44}$$

where, according to Theorems 5.3.1 and 5.3.2, the control applied to the system has been proved to be bounded by b_{u^*} at each ILC iteration, i.e.,

$$b_{u^*} \stackrel{\Delta}{=} \sup_{t\in[0,T]} \|u_i(t)\|, \forall i. \tag{5.45}$$

Denote $\Delta\bar{h}_i(t) \stackrel{\Delta}{=} \bar{h}_i(t) - \bar{h}_{i-1}(t), \quad \bar{h} \in \{u,y,w,v\}$. Similar to (5.10) and (5.11), the following two inequalities can be obtained

$$\Delta x_i(t+1) = \Delta f_i(t) + \Delta B_i(t)u_i(t) + B_i(t)\Delta u_i(t) + \Delta w_i(t), \tag{5.46}$$

$$\|\Delta x_i(t+1)\| \leq \tilde{e}\|\Delta x_i(t)\| + b_B\|\Delta u_i(t)\| + b'_w \tag{5.47}$$

where $b'_w \stackrel{\Delta}{=} \sup_{t\in[0,N]} \|\Delta w_i(t)\|, \forall i$. It is also apparent by referring to (5.18) that

$$\|\Delta x_i(t)\|_\lambda \leq b'_{x_0} + b_B \tilde{O}(|\lambda|^{-1})\|\Delta u_i(t)\|_\lambda + c'_0 b'_w \tag{5.48}$$

where $b'_{x_0} \stackrel{\Delta}{=} \|\Delta x_i(0)\|, \forall i$, and

$$\tilde{O}(|\lambda|^{-1}) \stackrel{\Delta}{=} \frac{1 - \tilde{e}^{-(\lambda-1)N}}{\tilde{e}^\lambda - \tilde{e}}, \quad c'_0 \stackrel{\Delta}{=} \sup_{t\in[0,N]} \frac{\tilde{e}^{-(\lambda-1)t}(1 - \tilde{e}^{-t})}{\tilde{e} - 1}.$$

The following notations are employed.

$$\bar{\rho}_1 \stackrel{\Delta}{=} \sup_{t\in[0,N]} \|I_r - C(t+1)B_i(t)Q_1(t)\|, \quad \forall i,$$

$$\bar{\rho}_k \stackrel{\Delta}{=} \sup_{t\in[0,N]} \|C(t+1)B_i(t)Q_k(t)\|, \quad \forall i, (k = 2,3,\cdots,M),$$

$$b'_v \stackrel{\Delta}{=} \sup_{t\in[0,N]} \|\Delta v_i(t)\|, \forall i.$$

To show that the tracking error bound is a function of the bounds of the differences of uncertainties, disturbances and the re-initialization errors between two successive ILC iterations even without the feedback controller, in addition to Assumptions A1) to A3), the following Assumptions A1') and A2') are imposed.

- A1'). The the differences of uncertainties, disturbances and the errors of re-initialization between two successive ILC iterations are bounded with unknown bounds denoted by b'_w, b'_v and b'_{x_0} respectively.
- A2'). Matrix $C(\cdot)B(\cdot,\cdot)$ has a full row rank $\forall t \in [0,N], x(t) \in R^n$.

According to Assumption A2'), the number of outputs r should be less than or equal to the number of inputs m. Because A2) is still applied for a bounded input sequence $\{u_i(t) \mid i = 1, 2, \cdots\}$, thus the restriction here is actually that $m = r$ and $C(\cdot)B(\cdot, \cdot)$ is full ranked.

Under these restrictions, new property of the tracking error convergence can be explored and is presented in the following theorem.

Theorem 5.3.3 *Given the realizable desired trajectory $y_d(t)$ over the fixed time interval $[0, NT_s]$, for the repetitive discrete-time uncertain time-varying nonlinear system (5.1) under assumptions A1)-A4), A1') and A2'), by using the high-order ILC updating law (5.9), the λ-norm of the tracking errors $e_i(t)$ is bounded if*

$$\sum_{k=1}^{M} \bar{\rho}_k < 1. \tag{5.49}$$

Moreover, the bound is a function of b'_w, b'_v and b'_{x_0}. Furthermore, the tracking error $e_i(t)$ converges uniformly to zero for $t = 1, \cdots, N$ as $i \to \infty$ when the uncertainties, disturbances and the re-initialization errors between two successive ILC iterations tend to be the same, i.e., when b'_w, b'_v and b'_{x_0} tend to 0.

Proof. The tracking error at $(i + 1)-$th ILC iteration can be expressed as

$$e_{i+1}(t+1) = e_i(t+1) - \Delta y_{i+1}(t+1)$$
$$= e_i(t+1) - C(t+1)\Delta x_{i+1}(t+1) - \Delta v_{i+1}(t+1). \tag{5.50}$$

Substituting (5.46) and (5.9) into (5.50) gives

$$e_{i+1}(t+1) = e_i(t+1) - C(t+1)[\Delta f_{i+1}(t) +$$
$$\Delta B_{i+1}(t)u_{i+1}(t)] - C(t+1)B_i(t)\sum_{k=1}^{M} Q_k(t)e_{i-k+1}(t+1) +$$
$$-C(t+1)\Delta w_{i+1}(t) - \Delta v_{i+1}(t+1). \tag{5.51}$$

Collecting terms and then taking the norm for (5.51) yield

$$\|e_{i+1}(t+1)\| \leq \sum_{k=1}^{M} \bar{\rho}_k\|e_{i-k+1}(t+1)\| + b_C\tilde{e}\|\Delta x_{i+1}(t)\|$$
$$+b_C b'_w + b'_v. \tag{5.52}$$

Taking the $\lambda-$norm for (5.52) and using the relationship (5.48), we simply have

$$\|e_{i+1}(t+1)\|_\lambda \leq \sum_{k=1}^{M} \bar{\rho}_k\|e_{i-k+1}(t+1)\|_\lambda + \varepsilon'$$
$$+ \sum_{k=1}^{M} \tilde{O}_k(|\lambda|^{-1})\|e_{i-k+1}(t+1)\|_\lambda \tag{5.53}$$

where $\tilde{O}_k(|\,\lambda\,|^{-1}) \triangleq b_{Q_k} b_C \bar{e} b_B \tilde{O}(|\,\lambda\,|^{-1})$ and $\varepsilon' \triangleq b_C \bar{e}(b'_{x_0} + c'_0 b'_w) + b_C b'_w + b'_v$.
Clearly, based on (5.49), a sufficiently large $|\,\lambda\,|$ can be used to ensure that

$$\sum_{k=1}^{M} [\bar{\rho}_k + \tilde{O}_k(|\,\lambda\,|^{-1})] < 1. \tag{5.54}$$

Therefore, according to Lemma 2.3.1, the λ-norm of tracking error is bounded
as $i \to \infty$. Referring to the definitions of ε' and $\tilde{O}_k(|\,\lambda\,|^{-1})$, we observe that
the tracking error bound is a function of b'_w, b'_v and b'_{x_0}. Furthermore, the
tracking error $e_i(t)$ converges uniformly to zero for $t = 1, 2, \cdots, N$ as $i \to \infty$
when the uncertainties, disturbances and the re-initialization errors between
two successive ILC iterations tend to be the same, i.e., when b'_w, b'_v and b'_{x_0}
tend to 0.

Remark 5.3.4. By referring to the Corollary 5.3.1, we know the Theorem
5.3.3 still holds if the ILC updating law (5.35) is applied.

Remark 5.3.5. Consider the case that the desired trajectory varies with respect to ILC iteration number i. Suppose the desired trajectory at the i-th
iteration is changed to $y_{d_i}(t)$. If

$$\|y_{d_{i+1}}(t) - y_{d_i}(t)\| < b'_{y_d}, \quad \forall t \in [0, N] \text{ and } \forall i, \tag{5.55}$$

then, all the conclusions made above are all valid by replacing b'_v with $b'_v + b'_{y_d}$.

Remark 5.3.6. It is implied in Theorem 5.3.3 that the ILC method can reject
any repetitive components of the uncertainty or disturbance. Uniform bound
of the tracking error is dependent on the bounds of the differences of the
uncertainties and disturbances between two successive system repetitions.

5.4 Simulation Illustrations

To demonstrate the effectiveness of the proposed feedback-assisted high-order
ILC algorithm for the improvement of the ILC convergence performance, a
single link direct joint driven manipulator model is used for the simulation
study. The dynamic equation in the continuous-time t' domain is

$$\ddot{\theta}(t') = \frac{1}{J}\tau(t') + \tau_n(t') + \frac{1}{J}(\frac{1}{2}m_0 + M_0)gl\sin\theta(t') \tag{5.56}$$

where $\theta(t')$ is the angular position of the manipulator; $\tau(t')$ is the applied
joint torque; $\tau_n(t')$ is the exogenous disturbance torque; m_0, l are the mass
and length of the manipulator respectively, M_0 is the mass of the tip load, g
is the gravitational acceleration and the J is the moment of inertia w.r.t the
joint, i.e., $J = M_0 l^2 + m_0 l^2/3$. The parameters used in this simulation study

are the same as in Table 2.2. Let the sampling period $T_s = 0.01$ sec. One can discretize the above model by using simple Euler method as follows:

$$\begin{cases} x_1(t+1) = x_2(t) \\ x_2(t+1) = 2x_2(t) - x_1(t) + [\tau(t) + \tau_n(t) \\ \qquad\qquad + (0.5m_0 + M_0)gl\sin(x_1(t))]T_s^2/J \end{cases} \tag{5.57}$$

where discrete time $t = 0, 1, \cdots, 100$, $x_1(t) = \theta(t), x_2(t) = \theta(t+1)$. The desired tracking trajectories of the tracking tasks over the time interval $[0,1]$ sec. are specified as

$$\theta_d(t) = \theta_b + (\theta_b - \theta_f)(15\tau_0^4 - 6\tau_0^5 - 10\tau_0^3) \tag{5.58}$$

where $\tau_0 = tT_s/(t_f - t_0)$. In the simulation, we use $\theta_b = 0°$, $\theta_f = 90°$, $t_0 = 0$, $t_f = 1$. The initial states at each ILC repetition are all set to 0. The ILC ends when $e_{b1} \leq 1°$ where

$$e_{b1} \overset{\triangle}{=} \sup_{t \in [0,100]} |\theta_d(t) - \theta(t)|.$$

To simplify the presentation, the following ILC updating law is used

$$u_{i+1}^f(t) = u_i(t) + \sum_{k=1}^{M} Q_k(e_{i-k+1}(t+1) - e_{i-k+1}(t)) \tag{5.59}$$

where Q_k is the learning parameter and $e(t) = \theta_d(t) - \theta(t)$. In the ILC method, the system parameters are assumed unknown. However, the learning parameter determination should be based on the knowledge of the system. To make the results with comparability, in the following simulations, we use $Q_1 = 50, Q_2 = 25, Q_3 = 25$.

The following six cases are presented. In cases 1-5, no control saturater is considered. The effect of a control saturater is shown in case 6.

- *Case 1. High-order Effects Without Feedback Controller*
  Consider the ideal situation without exogenous torque disturbance. The tracking control is achieved by the iterative learning controller only without the assistance of the feedback controller. To compare the effectiveness of the high-order ILC scheme, three subcases are considered, i.e., $M = 1, 2, 3$. The histories of the tracking error bounds are shown in Fig. 5.2 where the improved ILC convergence can be observed when higher order ILC updating law is applied.
- *Case 2. Effects of the Feedback Controller in ILC*
  This case is similar to *case 1*. We apply the fixed ILC updating law with $M = 1$. But a simple P-type feedback controller

$$u_i^b(t) = K_p e_i(t) \tag{5.60}$$

is applied to assist the ILC. Four subcases are considered for different gains K_p. When $K_p = 0$, the tracking is by ILC only which is the subcase $M = 1$ in *case 1*. The results are summarized in Fig. 5.3. It is observed

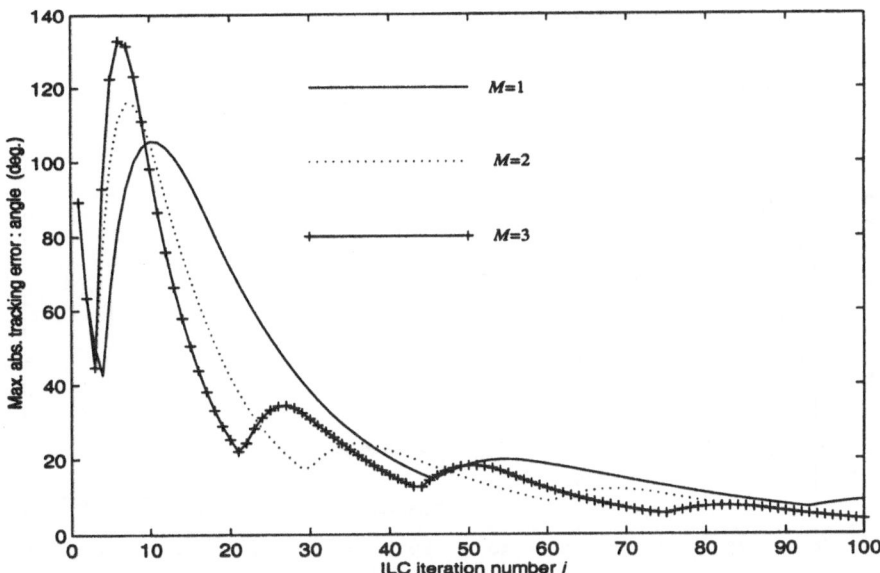

Fig. 5.2. Comparison of ILC convergence histories, ideal case with different ILC orders

that an improved ILC convergence performance can be obtained when the ILC is assisted by a feedback controller. It is interesting to note in this case that the ILC performance improves as K_p increases which implies a well designed feedback controller is quite helpful for the iterative learning control.

- *Case 3. High-order Effects With a Feedback Controller*
  Consider the ILC updating laws with different orders, i.e., $M = 1, 2, 3$. The same incorporated feedback controller is applied with $K_p = 30$. The tracking results are given in Fig. 5.4, where a similar performance improvement behavior of the ILC convergence can be observed.
- *Case 4. Tracking the Varying Desired Trajectories*
  In this case, we choose $M = 1$, $K_p = 30$. The desired trajectory is supposed to be able to vary with respect to ILC iteration number i as discussed in Remarks 5.3.1 and 5.3.5. The varying desired trajectories are designed with the same form (5.58) but the final angle $\theta_f(t)$ is revised to

$$\theta_{f_i}(t) = (2 - e^{-0.05i})45°. \tag{5.61}$$

The varying desired trajectory is denoted by θ_{d_i}. In this case, we choose $N = 1$, $K_p = 30$. In the first ILC iteration, the desired trajectory $\theta_{d_0}(t)$ and the system output $\theta_0(t)$ are plotted in Fig. 5.5. To have a clear comparison, the desired trajectory and the system output at 100-th ILC iteration, i.e., $\theta_{d_{100}}(t)$ and $\theta_{100}(t)$, are drawn in the same figure. We observe a good tracking in this case. It should be noted that in this case b_{y_d} defined

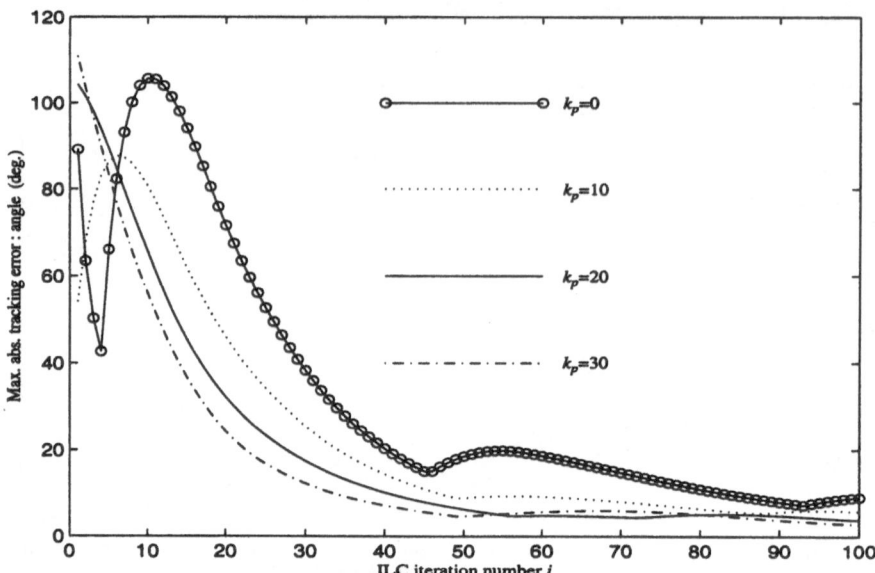

Fig. 5.3. Comparison of feedback-assisted ILC convergence histories, ideal case with the ILC order fixed

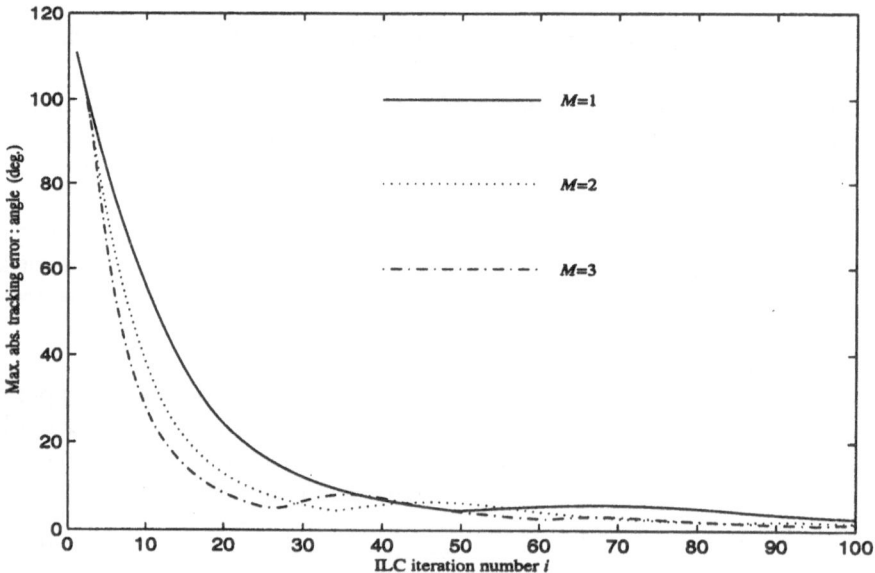

Fig. 5.4. Comparison of feedback-assisted ILC convergence histories, ideal case with different ILC orders

in Remark 5.3.1 is fairly large. However, according to Remark 5.3.5, the tracking error bound in this case should tend to 0 because b'_{y_d} tends to 0 as i increases. This is clearly illustrated by Fig. 5.5 which indicates that Remark 5.3.1 is more conservative than Remark 5.3.5.

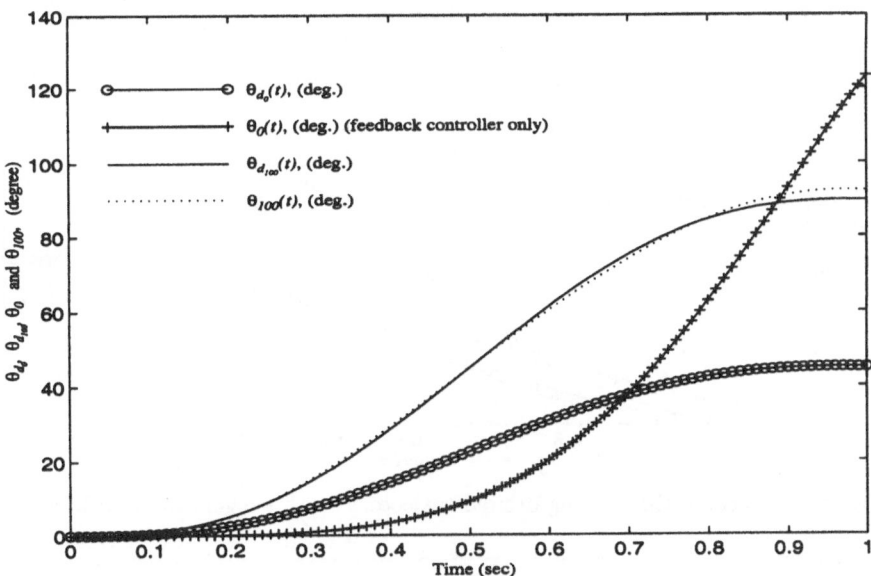

Fig. 5.5. Tracking the varying desired trajectory: ideal case

- *Case 5. Tracking the Varying Desired Trajectories With Exogenous Disturbance*
  The same situation as *case 4* is considered but with exogenous torque disturbance

$$\tau_n(t) = 70(2 - e^{-0.05i})\sin(\pi t/5) \quad \text{Nm}. \tag{5.62}$$

Good tracking performance in this case can be observed. It should be noted that in this case b_w defined in Theorem 5.3.1 is fairly large. However, according to Theorem 5.3.3, the tracking error bound in this case should tend to 0 because b'_w tends to 0 as i increases. This is clearly illustrated by Fig. 5.6 which indicates that the repetitive unknown exogenous disturbance can be rejected by ILC scheme.

- *Case 6. Effect of A Control Saturater*
  In case 2, the final converged control inputs are all the same for different feedback gains. The maximum amplitude of $u_d(t)$ is found to be 36 Nm. It is also found that, in some of the ILC iteration processes, the control $u_i(t)$ exceeds 36 Nm at some time instants. To illustrate that the ILC scheme analyzed in the above is still effective under control saturation, according to A4), the saturater limit is set as $\bar{u}^* = 36\text{Nm}$. Repeat the same simulation as in case 2. Similar results are obtained which are summarized in Fig. 5.7 and are consistent to the analysis of this chapter.

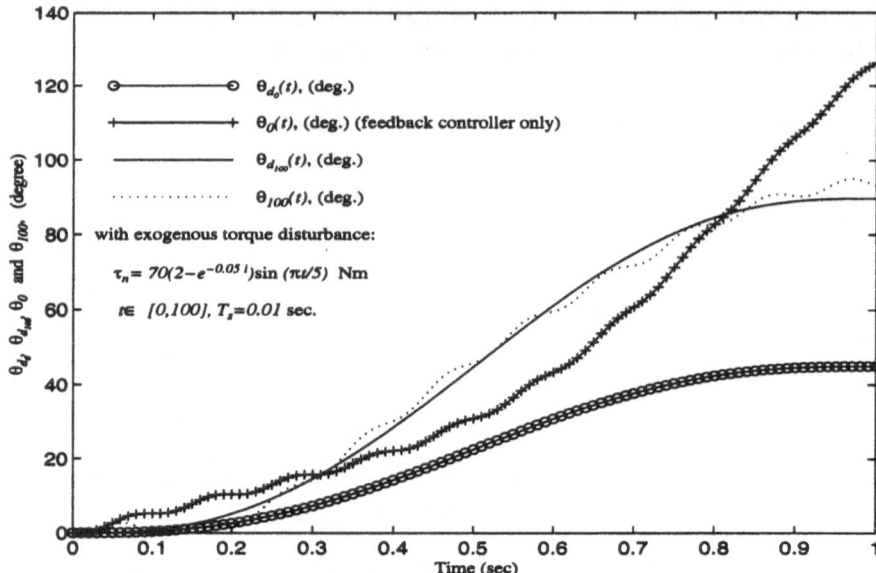

Fig. 5.6. Tracking the varying desired trajectory: with exogenous disturbance

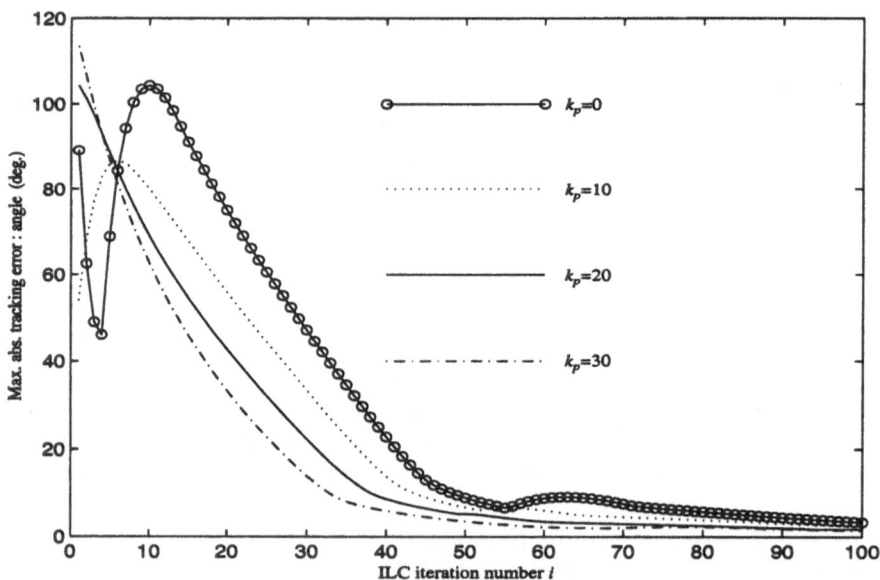

Fig. 5.7. Convergence history comparison for schemes of ILC plus feedback controller: with an input saturater

5.5 Conclusion

A high-order iterative learning controller, assisted by a feedback controller, is proposed. The effect of the control saturater is also considered. It has been proven that the tracking errors are bounded in the presence of bounded uncertainty, disturbance, and re-initialization error. Moreover, it has been shown that the tracking error bounds are functions of the bounds of uncertainty, disturbance, and re-initialization error. The tracking error bounds tend to 0 in the absence of uncertainty, disturbance, and re-initialization error. Furthermore, it also has been shown that under certain conditions, the tracking error bound is a function of the bounds of the differences of uncertainties, disturbances and the re-initialization errors between two successive ILC iterations. Improved ILC performances in both time t-axis and iteration number i-direction can be achieved by the proposed feedback-assisted high-order iterative learning controller. Simulation illustrations are presented to verify the effectiveness of the proposed schemes.

6. Initial State Learning Method for Iterative Learning Control of Uncertain Time-varying Systems

6.1 Introduction

IN CONVENTIONAL ITERATIVE LEARNING CONTROL (ILC), a common assumption is that the initial states in each repetitive operation should be inside a given ball centered at the desired initial states which may not be available. This assumption is critical to the ILC convergence analysis. All existing results on robustness analysis of ILC only guarantee the boundedness of the final tracking errors. These error bounds are not only directly related to the bounds of uncertainties and disturbances but also directly related to the initialization error bounds.

The concept of *closed-loop ILC* was proposed to employ the current cycle error in the ILC updating law [257, 132, 184]. With this, we do have a measure to reduce the effect of initialization error on the final tracking error bounds by increasing the gain of the current cycle error in the ILC updating law. This was achieved at the expense of using high gain control. Also, one critical question raised is how the first (initial) point iteratively learns because the ILC is in fact a *point-wise* scheme as explained in [109]. Under the assumption that the input transmission term appears in the system's output equation, for example the system model used for ILC convergence analysis [27, 225], an impulsive initial input [143] could be used to compensate the output tracking error so that it could finally approach to zero. But the use of an impulsive initial input is not practical.

Thus, how to totally eliminate the effect of the initialization errors on the final tracking error bounds is still an open problem. In this chapter, we will show that, the conventional assumption on re-initialization can be removed by using an initial state learning scheme together with a traditional D-type ILC updating law. Both linear and nonlinear time-varying uncertain systems are investigated. Uniform bounds for the final tracking errors are obtained and these bounds are only dependent on the system uncertainties and disturbances, yet independent of the initial errors. Furthermore, the desired initial states can be identified through learning iterations as illustrated in the simulations. The initial state learning method is extended to the case of high-order ILC updating law.

The remaining parts of this chapter are organized as follows. The first-order Initial State Learning scheme is proposed and analyzed in Sec. 6.2 and

Sec. 6.3 for linear and nonlinear time-varying uncertain systems, respectively. The effectiveness of the proposed initial state learning scheme is illustrated by some simulation results in Sec. 6.4. High-order initial state learning scheme is presented and analyzed in Sec. 6.5 with an illustrative example. Conclusions are made in Sec. 6.6.

6.2 Linear Time-varying Uncertain Systems

Consider a repetitive linear time-varying system with uncertainty and disturbance as follows

$$\begin{cases} \dot{x}_i(t) = A(t)x_i(t) + B(t)u_i(t) + w_i(t) \\ y_i(t) = C(t)x_i(t) + v_i(t) \end{cases} \tag{6.1}$$

where i denotes the i-th repetitive operation of the system; $x_i(t) \in R^n$, $u_i(t) \in R^m$, and $y_i(t) \in R^r$ are the state, control input, and output of the system, respectively; $w_i(t)$, $v_i(t)$ are uncertainties or disturbances to the system; $t \in [t_0, T] \subseteq [0, T]$ is the time and t_0, T are given; $A(t), B(t)$ and $C(t)$ are the time-varying matrices with appropriate dimensions.

Given the realizable desired output trajectory $y_d(t)$, the tracking error $e_i(t)$ at i-th repetition is that $e_i(t) \overset{\triangle}{=} y_d(t) - y_i(t)$. Then the problem is formulated as follows. Starting from an arbitrary continuous initial control input $u_0(t)$ and an arbitrary initial state $x_0(t_0)$, which may be different from $x_d(t_0)$, obtain the next control input $u_1(t)$ and initial state $x_1(t_0)$, and the subsequent series $\{u_i(t), x_i(t_0) \mid i = 2, 3, \cdots\}$ for system (6.1), in such a way that when $i \to \infty, y_i(t) \to y_d(t)$ and $x_i(t_0) \to x_d(t_0)$. Furthermore, $y_i(t) - y_d(t)$ and $x_i(t) - x_d(t)$ are independent of the initialization error $x_i(t_0) - x_d(t_0)$ when $i \to \infty$.

To solve the above problem, we shall use the D-type ILC updating law [17], i.e.,

$$u_{i+1}(t) = u_i(t) + L(t)\dot{e}_i(t), \tag{6.2}$$

where $L(t) \in R^{m \times r}$ is the continuous learning matrix, together with an initial state learning algorithm given to be

$$x_{i+1}(t_0) = x_i(t_0) + B(t_0)L(t_0)e_i(t_0). \tag{6.3}$$

To restrict our discussion, the following assumptions are made.

- A1). The uncertainty and disturbance terms $w_i(t)$ and $v_i(t)$ are bounded as follows, $\forall t \in [t_0, T]$ and $\forall i$,

$$\|w_{i+1}(t) - w_i(t)\|_\lambda \leq b_w, \quad \|v_{i+1}(t) - v_i(t)\|_\lambda \leq b_v,$$

- A2). For $t \in [t_0, T]$, matrices $B(t)$ and $C(t)B(t)$ have full column ranks.
- A3). $B(t)$ and $L(t)$ are differentiable over $[t_0, T]$. Furthermore, it is required that $L(t_0) \neq 0, \quad B(t_0) \neq 0$.

Assumption A1) puts the boundedness restrictions on the differences of the uncertainties and disturbances between two successive system repetitions. A3) is a reasonable assumption which makes the initial state correction possible. In this chapter, a common fundamental knowledge is that for a given bounded desired output $y_d(t)$, there exists a unique bounded input $u_d(t)$, $t \in [t_0, T]$ such that when $u(t) = u_d(t)$, the system has a unique bounded state $x_d(t)$ and $y_d(t) = C(t)x_d(t), t \in [t_0, T]$.

The general solution of state equation (6.1) can be written in the following form

$$x_i(t) = \Phi(t, t_0)\{x_i(t_0) + \int_{t_0}^{t} \Phi(t_0, \tau)[B(\tau)u_i(\tau) + w_i(\tau)]d\tau\} \qquad (6.4)$$

where $\Phi(t, t_0)$ stands for the state transition matrix of system (6.1). For brevity of our discussion, in the sequel, the following notations are used

$$\phi(t_0, t) \triangleq \frac{d}{dt}(\Phi(t_0, t)B(t)L(t)), \quad b_\phi \triangleq \sup_{t \in [t_0, T]} \|\phi(t_0, t)\|,$$

$$b_C \triangleq \sup_{t \in [t_0, T]} \|C(t)\|, \quad b_\Phi \triangleq \sup_{t \in [t_0, T]} \|\Phi(t, t_0)\|,$$

$$\varphi(t) \triangleq d(B(t)L(t))/dt, \quad b_\varphi \triangleq \sup_{t \in [t_0, T]} \|\varphi(t)\|,$$

$$b_{BL} \triangleq \sup_{t \in [t_0, T]} \|B(t)L(t)\|.$$

Theorem 6.2.1 *For the repetitive linear time-varying uncertain system (6.1) with assumptions A1)-A3), given the desired trajectory $y_d(t)$ over the fixed time interval $[t_0, T]$, by using the ILC updating law (6.2) and the initial state learning formula (6.3), then, the λ-norm of the output tracking error is bounded, if*

$$\|I_r - C(t)B(t)L(t)\| < 1, \quad \forall t \in [t_0, T]. \qquad (6.5)$$

For a sufficiently large λ,

$$\lim_{i \to \infty} \|e_i(t)\|_\lambda \leq \frac{b_v + O_1(\lambda^{-1})}{1 - \rho - O_2(\lambda^{-1})}, \quad \forall t \in [t_0, T], \qquad (6.6)$$

where

$$\rho \triangleq \sup_{t \in [t_0, T]} \|I_r - C(t)B(t)L(t)\|, \qquad (6.7)$$

$$O_1(\lambda^{-1}) \triangleq b_w b_C b_\Phi \|\Phi(t_0, t)\|_\lambda / \lambda, \qquad (6.8)$$

$$O_2(\lambda^{-1}) \triangleq b_C b_\Phi b_\phi / \lambda. \qquad (6.9)$$

In the case that the uncertainties and disturbances in the ILC iterations tend to be the same, i.e., $b_w \to 0$ and $b_v \to 0$, we have $e_i(t) \to 0$, i.e., $y_i(t) \to y_d(t)$, and also $x_i(t) \to x_d(t)$, $u_i(t) \to u_d(t)$ as $i \to \infty$ for all $t \in [t_0, T]$.

Proof. Using (6.4) together with (6.1), (6.2), and (6.3), the tracking error $e_{i+1}(t)$ can be expressed as

$$
\begin{aligned}
e_{i+1}(t) &= y_d(t) - y_{i+1}(t) \\
&= y_d(t) - v_{i+1}(t) - C(t)\Phi(t,t_0)x_i(t_0) \\
&\quad -C(t)\Phi(t,t_0)B(t_0)L(t_0)e_i(t_0) \\
&\quad -C(t)\Phi(t,t_0)\int_{t_0}^{t}\Phi(t_0,\tau)(B(\tau)u_i(\tau)+w_i(\tau))\mathrm{d}\tau \\
&\quad -C(t)\Phi(t,t_0)\int_{t_0}^{t}\Phi(t_0,\tau)B(\tau)L(\tau)\dot{e}_i(\tau)\mathrm{d}\tau \\
&\quad -C(t)\Phi(t,t_0)\int_{t_0}^{t}\Phi(t_0,\tau)(w_{i+1}(\tau)-w_i(\tau))\mathrm{d}\tau.
\end{aligned}
\tag{6.10}
$$

Integrating the term $\dot{e}_i(\cdot)$ in (6.10) by parts yields

$$
\begin{aligned}
-C(t)\Phi(t,t_0)\int_{t_0}^{t}\Phi(t_0,\tau)B(\tau)L(\tau)\dot{e}_i(\tau)\mathrm{d}\tau &= \\
-C(t)\Phi(t,t_0)\Phi(t_0,t)B(t)L(t)e_i(t) & \\
+C(t_0)\Phi(t,t_0)\Phi(t_0,t_0)B(t_0)L(t_0)e_i(t_0) & \\
+C(t)\Phi(t,t_0)\int_{t_0}^{t}[\tfrac{\mathrm{d}}{\mathrm{d}\tau}(\Phi(t_0,\tau)B(\tau)L(\tau))]e_i(\tau)\mathrm{d}\tau & \\
= -C(t)B(t)L(t)e_i(t) + C(t_0)\Phi(t,t_0)B(t_0)L(t_0)e_i(t_0) & \\
+C(t)\Phi(t,t_0)\int_{t_0}^{t}\phi(t_0,\tau)e_i(\tau)\mathrm{d}\tau. &
\end{aligned}
\tag{6.11}
$$

By substituting (6.11) to (6.10), we get

$$
\begin{aligned}
e_{i+1}(t) &= [I_r - C(t)B(t)L(t)]e_i(t) - (v_{i+1}(t) - v_i(t)) \\
&\quad -C(t)\Phi(t,t_0)\int_{t_0}^{t}\Phi(t_0,\tau)(w_{i+1}(\tau)-w_i(\tau))\mathrm{d}\tau \\
&\quad +C(t)\Phi(t,t_0)\int_{t_0}^{t}\phi(t_0,\tau)e_i(\tau)\mathrm{d}\tau.
\end{aligned}
\tag{6.12}
$$

Taking the norm of (6.12), we have

$$
\begin{aligned}
\|e_{i+1}(t)\| \leq \rho\|e_i(t)\| + b_v + b_w b_C b_\Phi \int_{t_0}^{t}\|\Phi(t_0,\tau)\|\mathrm{d}\tau \\
+ b_C b_\Phi b_\phi \int_{t_0}^{t}\|e_i(t)\|\mathrm{d}\tau
\end{aligned}
\tag{6.13}
$$

Multiplying $e^{-\lambda t}$ on both sides of (6.13) and then taking the λ-norm give

$$
\begin{aligned}
\|e_{i+1}(t)\|_\lambda \leq \rho\|e_i(t)\|_\lambda + b_v + b_w b_C b_\Phi\|\Phi(t_0,t)\|_\lambda O(\lambda^{-1}) \\
+ b_C b_\Phi b_\phi\|e_i(\tau)\|_\lambda O(\lambda^{-1}),
\end{aligned}
\tag{6.14}
$$

where

$$O(\lambda^{-1}) = \frac{1 - e^{-\lambda(t-t_0)}}{\lambda} \leq \frac{1}{\lambda}, \quad \forall t \in [t_0, T]. \tag{6.15}$$

Referring to (6.8) and (6.9), we can simply write (6.14) as

$$\|e_{i+1}(t)\|_\lambda \leq \bar{\rho}\|e_i(t)\|_\lambda + \varepsilon, \tag{6.16}$$

$$\|e_i(t)\|_\lambda \leq \bar{\rho}^i\|e_0(t)\|_\lambda + \frac{1 - \bar{\rho}^i}{1 - \bar{\rho}}\varepsilon, \tag{6.17}$$

where

$$\bar{\rho} = \rho + O_2(\lambda^{-1}), \tag{6.18}$$

$$\varepsilon = b_v + O_1(\lambda^{-1}). \tag{6.19}$$

Clearly, $\exists \lambda^* > 0$ such that $\bar{\rho} < 1$, $\forall \lambda \geq \lambda^*$. Thus

$$\lim_{i \to \infty} \|e_i(t)\|_\lambda \leq \frac{\varepsilon}{1 - \bar{\rho}} = \frac{b_v + O_1(\lambda^{-1})}{1 - \rho - O_2(\lambda^{-1})}. \tag{6.20}$$

When b_w and b_v tend to zero, as $i \to \infty$, $e_i(t) \to 0$, i.e., $y_i(t) \to y_d(t)$. Obviously, we also have $x_i(t) \to x_d(t)$, and $u_i(t) \to u_d(t)$ $\forall t \in [t_0, T]$ as $i \to \infty$. This completes the proof of Theorem 6.2.1.

Remark 6.2.1. The assumption A1) is less restrictive than the conventionally proposed one such as in [215]. In the case that at every ILC iteration the uncertainty and disturbance are all the same, i.e., they are repeatable, the final tracking error bound will be zero.

From (6.6), it can be seen that the initialization error has no effect on the final tracking error bound through the initial state learning scheme given in (6.3) together with the D-type ILC updating law (6.2). This property still holds for nonlinear systems by using the same ILC updating law (6.2) and initial state learning scheme (6.3).

6.3 Nonlinear Time-varying Uncertain Systems

The repetitive nonlinear time-varying uncertain system is described by

$$\begin{cases} \dot{x}_i(t) = f(x_i(t), t) + B(t)u_i(t) + w_i(t) \\ y_i(t) = C(t)x_i(t) + v_i(t). \end{cases} \tag{6.21}$$

Now, with the same assumptions, notations, and definitions as in Sec. 6.3 if not otherwise indicated, we intend to show that with the same ILC updating law (6.2) and initial state learning scheme (6.3), similar conclusion can be made for the above nonlinear time-varying uncertain system (6.21). Before presenting Theorem 6.2.1, we need one more assumption

- A4). $f(\cdot, \cdot) : R^n \times [t_0, T] \mapsto R^n$ is a piecewise continuous function and satisfies a Lipschitz continuity condition, i.e., $\forall t \in [t_0, T]$,

$$\|f(x_{i+1}(t), t) - f(x_i(t), t)\| \le k_f \|x_{i+1}(t) - x_i(t)\|,$$

where $k_f > 0$ is the Lipschitz constant.

Theorem 6.3.1 *For the repetitive nonlinear time-varying uncertain system (6.21) with assumptions A1)-A4), given the desired trajectory $y_d(t)$ over the fixed time interval $[t_0, T]$, by using the ILC updating law (6.2) and the initial state learning scheme (6.3), if (6.5) is satisfied, then the λ-norm of the output tracking error is bounded. For a sufficiently large λ,*

$$\lim_{i \to \infty} \|e_i(t)\|_\lambda \le \frac{2b_v + 2b_C b_w T + O_3(\lambda^{-1})}{1 - \rho - O_4(\lambda^{-1})}, \quad \forall t \in [t_0, T], \tag{6.22}$$

where

$$O_3(\lambda^{-1}) = \frac{b_C b_w k_f T O(\lambda^{-1})}{1 - k_f O(\lambda^{-1})}, \tag{6.23}$$

$$O_4(\lambda^{-1}) = b_C b_\varphi O(\lambda^{-1}) + \frac{b_C k_f O(\lambda^{-1})(b_\varphi O(\lambda^{-1}) + b_{BL})}{1 - k_f O(\lambda^{-1})}. \tag{6.24}$$

If the uncertainties and disturbances of successive ILC iterations tend to be the same, i.e., $b_w \to 0$ and $b_v \to 0$, we have $e_i(t) \to 0$, i.e., $y_i(t) \to y_d(t)$, and also $x_i(t) \to x_d(t)$, $u_i(t) \to u_d(t)$ as $i \to \infty$ for all $t \in [t_0, T]$.

Proof. The idea is similar to that of the proof of Theorem 6.2.1. The formula for the tracking error at $(i+1)-$th repetition is

$$
\begin{aligned}
e_{i+1}(t) &= e_i(t) - (y_{i+1} - y_i(t)) \\
&= e_i(t) - C(t)(x_{i+1} - x_i(t)) - (v_{i+1} - v_i(t))
\end{aligned}
\tag{6.25}
$$

Integrating (6.21) gives

$$
\begin{aligned}
x_{i+1}(t) - x_i(t) &= B(t_0)L(t_0)e_i(t_0) \\
&\quad + \int_{t_0}^t (f(x_{i+1}(\tau), \tau) - f(x_i(\tau), \tau))d\tau \\
&\quad + \int_{t_0}^t (w_{i+1}(\tau) - w_i(\tau))d\tau + \int_{t_0}^t B(\tau)L(\tau)\dot{e}_i(\tau)d\tau \\
&= B(t)L(t)e_i(t) \\
&\quad + \int_{t_0}^t (f(x_{i+1}(\tau), \tau) - f(x_i(\tau), \tau))d\tau \\
&\quad + \int_{t_0}^t (w_{i+1}(\tau) - w_i(\tau))d\tau - \int_{t_0}^t \varphi(\tau)e_i(\tau)d\tau.
\end{aligned}
\tag{6.26}
$$

Substituting (6.26) into (6.25) and taking norm yield

$$\|e_{i+1}(t)\| \le \rho\|e_i(t)\| + b_v + b_C b_w T +$$
$$b_C k_f \int_{t_0}^{t} \|x_{i+1}(\tau) - x_i(\tau)\| d\tau + b_C b_\varphi \int_{t_0}^{t} \|e_i(\tau)\| d\tau. \tag{6.27}$$

Taking the λ–norm for (6.27), we have

$$\|e_{i+1}(t)\|_\lambda \le \rho\|e_i(t)\|_\lambda + b_v + b_C b_w T$$
$$+ b_C k_f \|x_{i+1}(t) - x_i(t)\|_\lambda O(\lambda^{-1})$$
$$+ b_C b_\varphi \|e_i(t)\|_\lambda O(\lambda^{-1}). \tag{6.28}$$

Taking the λ–norm for (6.26) and assuming that λ is large enough to ensure

$$\lambda > k_f(1 - e^{-\lambda T}), \tag{6.29}$$

the relationship between $\|x_{i+1}(t) - x_i(t)\|_\lambda$ and $\|e_i(t)\|_\lambda$ is given by

$$\|x_{i+1}(t) - x_i(t)\|_\lambda \le \frac{b_w T + (b_\varphi O(\lambda^{-1}) + b_{BL})\|e_i(t)\|_\lambda}{1 - k_f O(\lambda^{-1})}. \tag{6.30}$$

By substituting (6.30) into (6.28), then $\|e_{i+1}(t)\|_\lambda$ can be expressed simply as

$$\|e_{i+1}(t)\|_\lambda \le \tilde{\rho}\|e_i(t)\|_\lambda + \tilde{\varepsilon}, \tag{6.31}$$

where

$$\tilde{\rho} = \rho + O_4(\lambda^{-1}) \tag{6.32}$$
$$\tilde{\varepsilon} = b_v + b_C b_w T + O_3(\lambda^{-1}). \tag{6.33}$$

Obviously, a sufficiently large λ that satisfies (6.22), (6.29), and the condition $\tilde{\rho} < 1$ simultaneously exists. This completes the proof by referring to the proof of Theorem 6.2.1.

6.4 Simulation Illustrations

The following uncertain time-varying nonlinear system is used for the simulation studies.

$$\begin{cases} \dot{X}_i(t) = A(X_i(t), t)X_i(t) + U_i(t) + W_i(t) \\ \begin{bmatrix} y_{1_i}(t) \\ y_{2_i}(t) \end{bmatrix} = \begin{bmatrix} 4x_{1_i}(t) \\ x_{2_i}(t) \end{bmatrix} + \begin{bmatrix} v_{1_i}(t) \\ v_{2_i}(t) \end{bmatrix} \end{cases}$$

where i is the system repetition number and the time $t \in [0,1]$, $X_i(t) = [x_{1_i}(t), x_{2_i}(t)]^T$, $U_i(t) = [u_{1_i}(t), u_{2_i}(t)]^T$, $W_i(t) = [w_{1_i}(t), w_{2_i}(t)]^T$, and

$$A(X_i(t), t) = \begin{bmatrix} \alpha_1 \sin(x_{2_i}(t)) & 1 + \alpha_1 \sin(x_{1_i}(t)) \\ -2 - 5t & -3 - 2t \end{bmatrix}.$$

The uncertainties and output disturbances are

$$\begin{bmatrix} w_{1_i}(t) \\ w_{2_i}(t) \end{bmatrix} \stackrel{\triangle}{=} \alpha_2 \begin{bmatrix} \cos(2\pi f_0 t) \\ 2\cos(4\pi f_0 t) \end{bmatrix},$$

$$\begin{bmatrix} v_{1_i}(t) \\ v_{2_i}(t) \end{bmatrix} \stackrel{\triangle}{=} \alpha_2 \begin{bmatrix} \sin(2\pi f_0 t) \\ 2\sin(4\pi f_0 t) \end{bmatrix}.$$

where $f_0 = 1/(20h)$ Hz.. The RK-4 method is used to numerically integrate the state equation with a fixed time step $h = 0.01$ second. The desired tracking trajectories $y_{1_d}(t) = y_{2_d}(t) \stackrel{\triangle}{=} 12t^2(1-t)$. Referring to equations (6.1)-(6.3), we know B=diag[1, 2] and C=diag[4, 1]. So, the best learning matrix $L^* = (CB)^{-1}$ =diag[0.25, 0.5]. However, the system dynamics is assumed unknown. It is reasonable to use the ILC updating law (6.2) with a modified learning gain matrix $L = \alpha_3 L^*$. And also in the initial state learning scheme (6.3), the best learning coefficient matrix BL^* =diag[0.25, 1] is replaced with a modified one $BL = \alpha_4 BL^*$. The coefficients α_3 and α_4 are freely chosen to accommodate the inaccurate knowledge of B and C. Without loss of generality, here we assume that at the first ILC iteration, the initial states are $x_{1_1}(0) = \alpha_5, x_{2_1}(0) = -\alpha_5$. Let $\alpha = [\alpha_1, \alpha_2, \cdots, \alpha_5]$. Clearly, when $\alpha = [0, 0, 1, 0, 0]$, the system reduces to the one considered in [17, 27]. In our simulation studies, the following three cases with different α's were examined.
Case 1: $\alpha = [1, \ 0, \ 1, \ 0, \ 0]$. This is an ideal case without any initialization error, uncertainty, and disturbance.
Case 2: $\alpha = [1, \ 1, \ 0.5, \ 0.5, \ 0.5]$. This implies an initialization error exists and an initial state learning scheme is applied. Some uncertainties also exist in B and C.
Case 3: $\alpha = [1, \ 1, \ 0.5, \ 0, \ 0.5]$. The initial state learning scheme is switched off. The amplitudes of disturbance and uncertainty are twice of the initialization error, which are the same as in **Case 2**.

    Let the final tracking error bound be $b_{e_{j_i}} \stackrel{\triangle}{=} \sup_{t\in[0,T]} | e_{j_i}(t) |$, $j = 1, 2$. Then the simulation results are shown in Figs. 6.1 to 6.3. For all the three cases, the ILC termination conditions are all the same, i.e, the simulation will stop if $b_{e_{j_i}} < 0.01$, $\forall j = 1, 2$.
    From Figs. 6.1 to 6.3, we can observe that the initial state learning scheme is effective. The initial states finally track the desired ones. In **Case 3**, we note that the final tracking error bounds are directly contributed by initialization errors. This can be explained from Remark 6.2.1 because $b_v = b_w = 0$.

6.5 High-order Initial State Learning

6.5.1 High-order Updating Law

As discussed before, a high-order ILC given below is capable of giving better performance

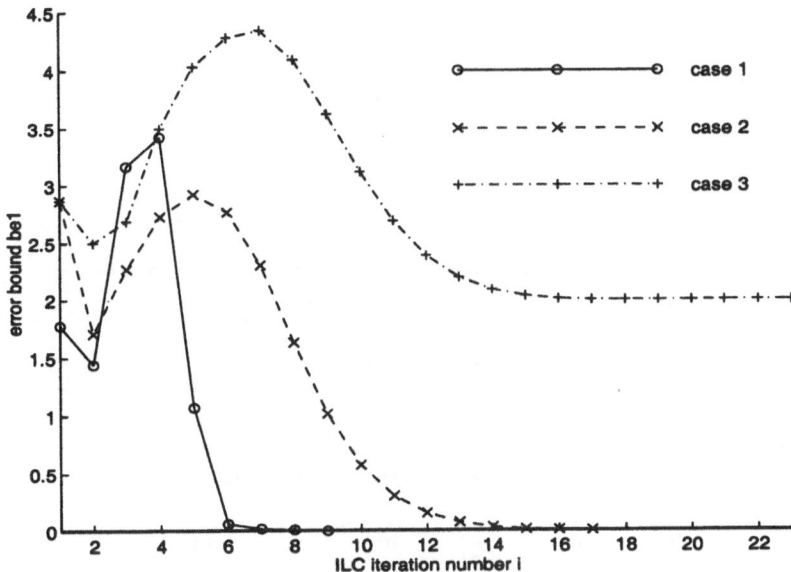

Fig. 6.1. The ILC histories of the output tracking error bounds $b_{e_{1_i}}$

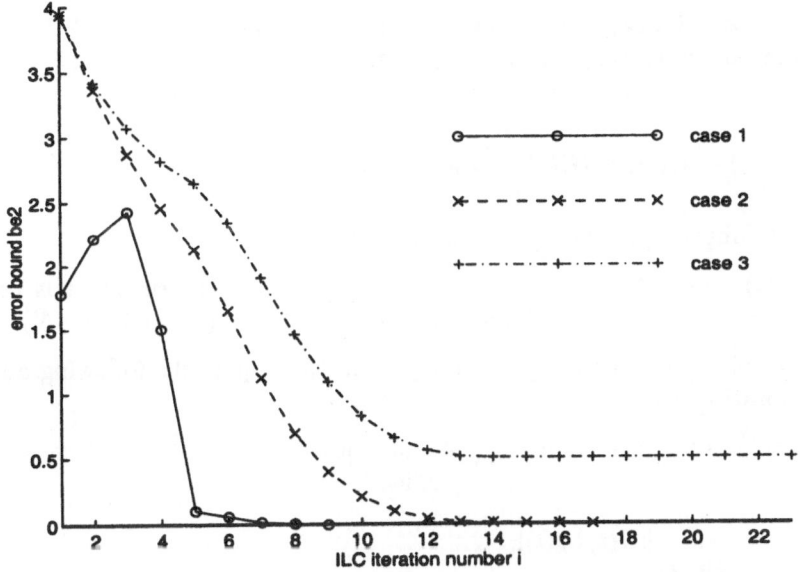

Fig. 6.2. The ILC histories of the output tracking error bounds $b_{e_{2_i}}$

$$u_{i+1}(t) = u_i(t) + \sum_{k=1}^{N} L_k(t)\dot{e}_{i-k+1}(t), \qquad (6.34)$$

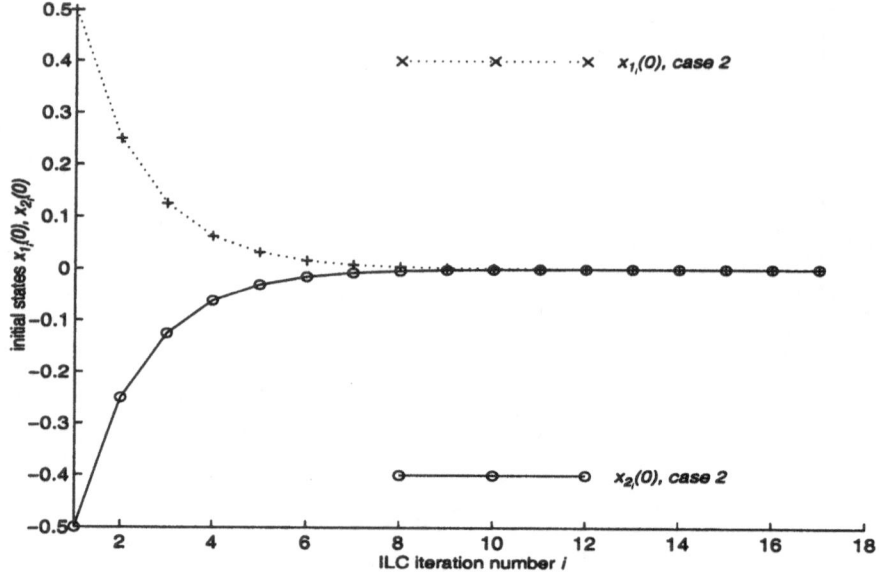

Fig. 6.3. The initial state learning processes $x_{1_i}(0)$ and $x_{2_i}(0)$

where the positive integer N is the order ILC updating law; $L_k(t) \in R^{m \times r}$ is the continuous learning matrix, together with a high-order initial state learning algorithm given to be

$$x_{i+1}(t_0) = x_i(t_0) + B(t_0) \sum_{k=1}^{N} L_k(t_0) e_{i-k+1}(t_0). \qquad (6.35)$$

The Assumption A3) is now extended as follows:

- A3'). $B(t)$ and $L_k(t)$ are differentiable over $[t_0, T]$. Furthermore, it is required that $L_k(t_0) \neq 0$, $B(t_0) \neq 0$ and $v_i(t_0) = 0$, $\forall i.$ $(k = 1, 2, \cdots, N)$

For brevity of the ILC convergence analysis, in the sequel, the following additional notations are used

$$\varphi_k(t) \stackrel{\Delta}{=} d(B(t)L_k(t))/dt, \quad b_{\varphi_k} \stackrel{\Delta}{=} \sup_{t \in [t_0, T]} \|\varphi_k(t)\|,$$

$$b_{BL_k} \stackrel{\Delta}{=} \sup_{t \in [t_0, T]} \|B(t)L_k(t)\|,$$

6.5.2 Convergence Analysis

The robust convergence property for the proposed high-order ILC with a high-order initial state learning scheme is presented in the following theorem.

Theorem 6.5.1 *Consider the repetitive nonlinear time-varying uncertain system (6.21) satisfying assumptions A1), A2), A3') and A4). For a desired trajectory $y_d(t)$ over the fixed time interval $[t_0, T]$, the ILC updating law (6.34) and the initial state learning scheme (6.35) are used. If*

$$\rho \triangleq \sum_{k=1}^{N} \rho_k < 1, \tag{6.36}$$

where

$$\rho_1 \triangleq \sup_{t \in [t_0, T]} \|I_r - C(t)B(t)L_1(t)\|,$$

$$\rho_k \triangleq \sup_{t \in [t_0, T]} \|C(t)B(t)L_k(t)\|, \quad k = 2, 3, \cdots, N$$

is satisfied, then the λ-norm of the output tracking error is bounded. For a sufficiently large λ,

$$\lim_{i \to \infty} \|e_i(t)\|_\lambda \le \frac{b_v + b_C b_w T + O_1(\lambda^{-1})}{1 - \sum_{k=1}^{N}(\rho_k + O_{2_k}(\lambda^{-1}))}, \quad \forall t \in [t_0, T], \tag{6.37}$$

where

$$O_1(\lambda^{-1}) = \frac{b_C b_w k_f T O(\lambda^{-1})}{1 - k_f O(\lambda^{-1})}, \tag{6.38}$$

$$O_{2_k}(\lambda^{-1}) = b_C b_{\varphi_k} O(\lambda^{-1}) + \frac{b_C k_f O(\lambda^{-1})(b_{\varphi_k} O(\lambda^{-1}) + b_{BL_k})}{1 - k_f O(\lambda^{-1})}. \tag{6.39}$$

Proof. The tracking error at $(i+1)$-th repetition is

$$e_{i+1}(t) = e_i(t) - (y_{i+1}(t) - y_i(t))$$
$$= e_i(t) - C(t)(x_{i+1}(t) - x_i(t)) - (v_{i+1}(t) - v_i(t)) \tag{6.40}$$

Integrating (6.21) gives

$$x_{i+1}(t) - x_i(t) = x_{i+1}(t_0) - x_i(t_0) + \int_{t_0}^{t} (f(x_{i+1}(\tau), \tau) - f(x_i(\tau), \tau)) d\tau$$

$$+ \int_{t_0}^{t} (w_{i+1}(\tau) - w_i(\tau)) d\tau + \int_{t_0}^{t} B(\tau)(u_{i+1}(\tau) - u_i(\tau)) d\tau$$

$$= B(t_0) \sum_{k=1}^{N} L_k(t_0) e_{i-k+1}(t_0) + \int_{t_0}^{t} (f(x_{i+1}(\tau), \tau) - f(x_i(\tau), \tau)) d\tau$$

$$+ \int_{t_0}^{t} (w_{i+1}(\tau) - w_i(\tau)) d\tau + \int_{t_0}^{t} B(\tau) \sum_{k=1}^{N} L_k(\tau) \dot{e}_{i-k+1}(\tau) d\tau$$

$$= B(t) \sum_{k=1}^{N} L_k(t) e_{i-k+1}(t) + \int_{t_0}^{t} (f(x_{i+1}(\tau), \tau) - f(x_i(\tau), \tau)) d\tau$$

$$+ \int_{t_0}^{t} (w_{i+1}(\tau) - w_i(\tau))\mathrm{d}\tau - \sum_{k=1}^{N} \int_{t_0}^{t} \varphi_k(\tau)e_{i-k+1}(\tau)\mathrm{d}\tau. \qquad (6.41)$$

Substituting (6.41) into (6.40) and taking norm yield

$$\|e_{i+1}(t)\| \le \sum_{k=1}^{N} \rho_k \|e_{i-k+1}(t)\| + b_v + b_C b_w T$$

$$+ b_C k_f \int_{t_0}^{t} \|x_{i+1}(\tau) - x_i(\tau)\|\mathrm{d}\tau$$

$$+ b_C \sum_{k=1}^{N} b_{\varphi_k} \int_{t_0}^{t} \|e_{i-k+1}(\tau)\|\mathrm{d}\tau. \qquad (6.42)$$

Taking the $\lambda-$norm for (6.42), we have

$$\|e_{i+1}(t)\|_\lambda \le \sum_{k=1}^{N} \rho_k \|e_{i-k+1}(t)\|_\lambda + b_v + b_C b_w T$$

$$+ b_C k_f \|x_{i+1}(t) - x_i(t)\|_\lambda O(\lambda^{-1}) + b_C \sum_{k=1}^{N} b_{\varphi_k} \|e_{i-k+1}(t)\|_\lambda O(\lambda^{-1}) (6.43)$$

By taking the $\lambda-$norm for (6.41) and assuming that λ is large enough to ensure

$$k_f O(\lambda^{-1}) < 1, \qquad (6.44)$$

the relationship between $\|x_{i+1}(t) - x_i(t)\|_\lambda$ and $\|e_{i-k+1}(t)\|_\lambda$ $(k = 1, 2, \cdots, N)$ is given by

$$\|x_{i+1}(t) - x_i(t)\|_\lambda \le \frac{b_w T + \sum_{k=1}^{N}(b_{\varphi_k} O(\lambda^{-1}) + b_{BL_k})\|e_{i-k+1}(t)\|_\lambda}{1 - k_f O(\lambda^{-1})}. (6.45)$$

By substituting (6.45) into (6.43), then $\|e_{i+1}(t)\|_\lambda$ can be expressed simply as

$$\|e_{i+1}(t)\|_\lambda \le \sum_{k=1}^{N} \bar{\rho}_k \|e_{i-k+1}(t)\|_\lambda + \bar{\varepsilon}, \qquad (6.46)$$

where

$$\bar{\rho}_k = \rho_k + O_{2_k}(\lambda^{-1}) \qquad (6.47)$$
$$\bar{\varepsilon} = b_v + b_C b_w T + O_1(\lambda^{-1}). \qquad (6.48)$$

Obviously, a sufficiently large λ that satisfies (6.37), (6.44), and the condition $\bar{\rho} < 1$ simultaneously exists. By applying Lemma 2.3.1, (6.37) is proved.

Corollary 6.5.1 *If the uncertainties and disturbances of successive ILC iterations tend to be the same, i.e., $b_w \to 0$ and $b_v \to 0$, we have $e_i(t) \to 0$, i.e., $y_i(t) \to y_d(t)$, and also $x_i(t) \to x_d(t)$, $u_i(t) \to u_d(t)$ as $i \to \infty$ for all $t \in [t_0, T]$.*

Proof. Referring to A3'), (6.37), (6.38), and (6.39), it is easy to observe that if the uncertainties and disturbances of successive ILC iterations tend to be the same, i.e., $b_w \to 0$ and $b_v \to 0$, we have $e_i(t) \to 0$, i.e., $y_i(t) \to y_d(t)$, and also $x_i(t) \to x_d(t)$, $u_i(t) \to u_d(t)$ as $i \to \infty$ for all $t \in [t_0, T]$.

Remark 6.5.1. A larger N may give a better ILC performance due to the increased flexibilities in choosing learning parameters at the expense of introducing more design parameters. The high-order ILC updating law can be regarded as the PID controller in the i-direction where the higher order information is used to approximate the derivative information in the i-direction.

Remark 6.5.2. Consider the case that the desired trajectory varies with respect to ILC iteration number i. Suppose the desired trajectory at i-th iteration is changed to $y_{d_i}(t)$. If

$$\|y_{d_{i+1}}(t) - y_{d_i}(t)\| < b_{y_d}, \quad \forall t \in [t_0, T] \text{ and } \forall i, \tag{6.49}$$

then, all the discussions made above are all valid by replacing b_v with $b_v + b_{y_d}$.

Remark 6.5.3. It is implied in Corollary 6.5.1 that the ILC method can reject any repetitive uncertainty or disturbance. Uniform bounds of the tracking errors are only dependent on the bounds of the differences of the uncertainties and disturbances between two successive system repetitions.

6.5.3 An Illustrative Example

Consider the same repetitive nonlinear system as in Sec. 6.4. Replace α_1 with 1 which results in the case of nonlinear system. The other conditions such as the RK-4 integration method, step size etc. are all the same. However, we re-define α_1 as the magnitude of the uncertainties and output disturbances which is given by

$$\begin{bmatrix} w_{1_i}(t) \\ w_{2_i}(t) \end{bmatrix} \triangleq \alpha_1 \begin{bmatrix} \cos(2\pi f_0 t) \\ 2\cos(4\pi f_0 t) \end{bmatrix}, \quad \begin{bmatrix} v_{1_i}(t) \\ v_{2_i}(t) \end{bmatrix} \triangleq \alpha_1 \begin{bmatrix} \sin(2\pi f_0 t) \\ 2\sin(4\pi f_0 t) \end{bmatrix}.$$

Other coefficients $\alpha_2, \cdots, \alpha_5$ are defined differently as follows. In the cases when $N = 1$ and $N = 2$, we can use the ILC updating law (6.34) with the modified learning gain matrices $L_1 = \alpha_2 L^*$ and $L_2 = \alpha_3 L^*$ under the constraint that $|1-\alpha_2|+|\alpha_3| < 1$. In (6.35), the matrix $B(t_0)$ is replaced with a modified $\alpha_4 B(t_0)$. Thus the initial state learning scheme can be switched off if α_4 is set to 0. To simplify our discussion, without loss of generality, we assume that at the first ILC iteration, the initial states are $x_{1_1}(0) = \alpha_5, x_{2_1}(0) = -\alpha_5$. It should be pointed out that the coefficients α_2, α_3, and α_4 are introduced to express the inaccurate knowledge of B and C. Proper choices of α_2, α_3, and α_4 to satisfy the ILC convergence condition are required.

To simplify the presentation of the simulation results, let

$$\alpha = [\alpha_1, \alpha_2, \cdots, \alpha_5].$$

The following 4 cases were examined which cover a wide range of different situations.

- Case 1: $\alpha = [1, 0.6, 0, 0, 0.5]$. This implies a re-initialization error exists and no initial state learning scheme is applied with ILC order $N = 1$. Some uncertainties also exist in B and C.
- Case 2: $\alpha = [1, 0.6, 0, 0.5, 0.5]$. This is the same as Case 1, but the first-order initial state learning scheme is applied.
- Case 3: $\alpha = [1, 0.6, 0.1, 0.5, 0.5]$. This is the same as Case 2, but the ILC order $N = 2$ and with $L_2 = 0.1L^*$.
- Case 4: $\alpha = [1, 0.6, -0.1, 0.5, 0.5]$. This is the same as Case 3, but with $L_2 = -0.1L^*$.

Let the final tracking error bound be denoted as

$$b_{e_{ji}} = \sup_{t \in [0,T]} |e_{ji}(t)|, \quad j = 1, 2.$$

Then the simulation results are shown in Figs. 6.4 - 6.6. For all the 4 cases, the ILC termination conditions are of the same, i.e, the simulation will stop if $b_{e_{ji}} < 0.01, \quad \forall j = 1, 2$.

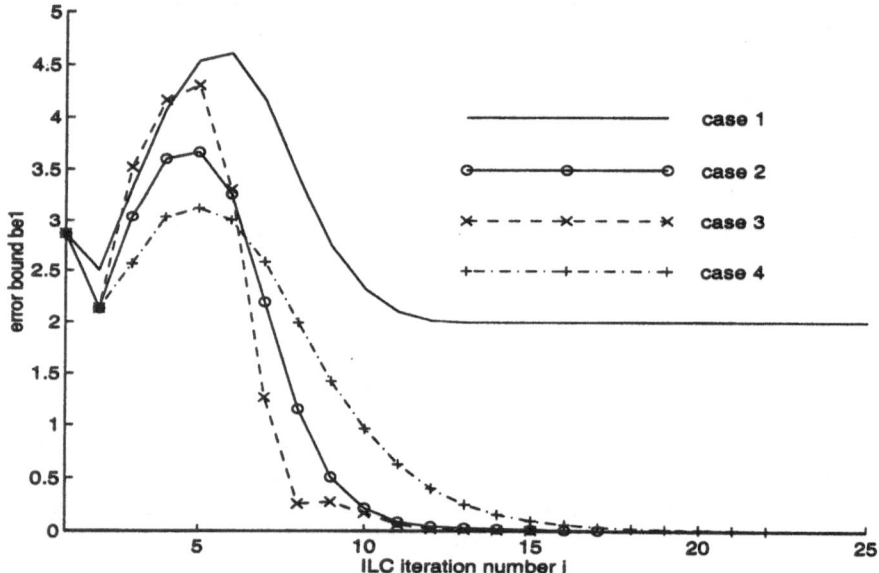

Fig. 6.4. The ILC histories of the tracking error bound e_{b_1}, high-order case

From Figs. 6.4-6.6, we can observe that the proposed high-order initial state learning scheme is effective. The initial states finally track the desired ones. The numbers of ILC iterations of Cases 2 to 4 are 16, 14, and 21 respectively. In Case 3 ($N = 2$), the ILC convergence is faster than that of Case 2 ($N = 1$) but Case 3 exhibits a larger overshoot. Compared to Case 2, the overshoot is reduced in Case 4 with relatively more ILC iterations.

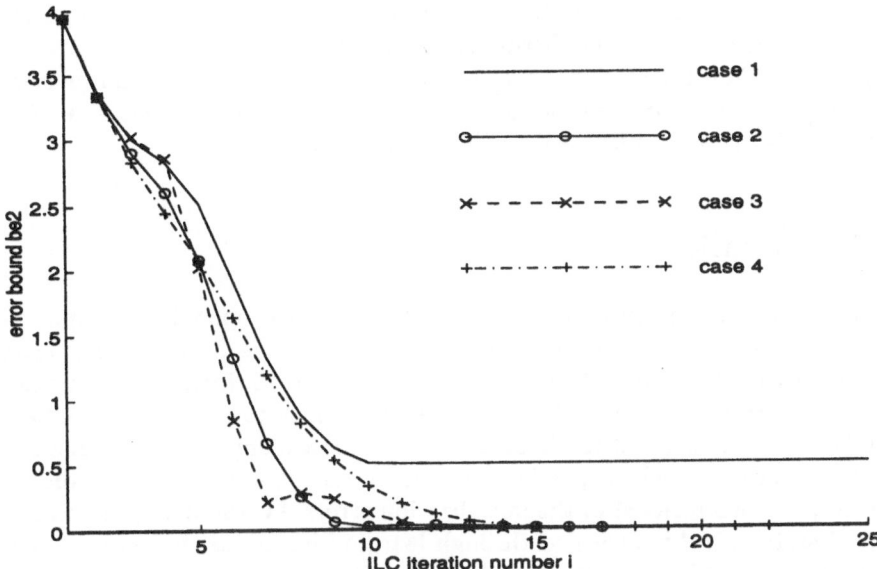

Fig. 6.5. The ILC histories of the tracking error bound e_{b_2}, high-order case

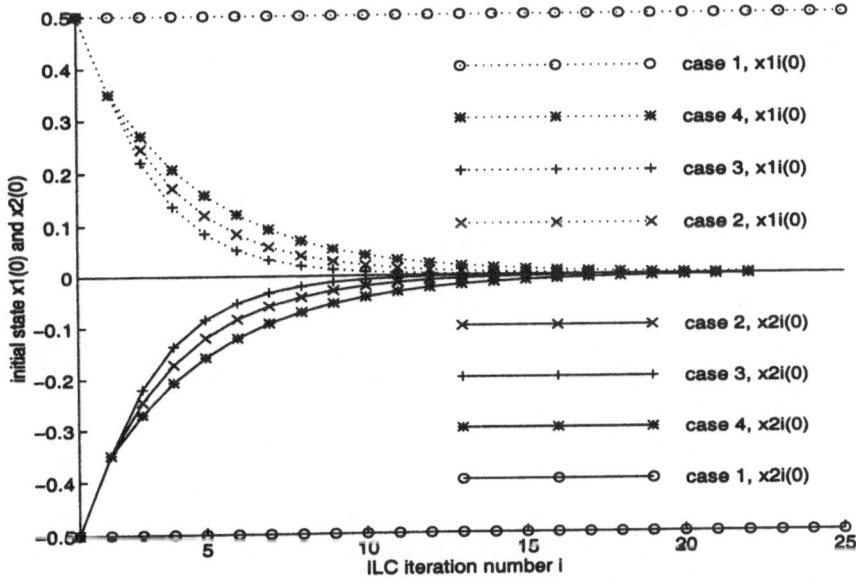

Fig. 6.6. Comparison of the initial state learning processes, high-order case

It is clear that the high-order scheme proposed in this section provides the capability to improve the ILC transient process along the ILC iteration number i-direction. Moreover, in all cases, we note that the final tracking error bounds are directly contributed by re-initialization errors even though the system has uncertainty and disturbance. This is because $b_v = b_w = 0$ according to A1). Thus the tracking error bounds tend to zero when the initial state learning scheme is applied for cases 2-4, which are shown in Figs. 6.4 - 6.5. This can be explained by (6.37) of Theorem 6.5.1, too.

6.6 Conclusion

In this chapter, an initial state learning scheme is proposed to completely eliminate the effect of the initialization errors on the final tracking error bounds through iterative learning control with a traditional D-type updating law. Both linear and nonlinear time-varying uncertain systems have been studied. It is shown that the final tracking errors are uniformly bounded and these bounds are only dependent on the system uncertainties and disturbances, but independent of the initialization errors. Furthermore, the desired initial states can be identified through learning iterations. It has also been clearly illustrated that, with a high-order initial state learning scheme, the effect of the re-initialization errors on the final tracking error bounds can be eliminated more effectively. The simulation results indicate that a high-order scheme may give an improved ILC transient performance along the ILC iteration number direction.

7. High-order Terminal Iterative Learning Control with an Application to a Rapid Thermal Process for Chemical Vapor Deposition

7.1 Introduction

INITIALLY, Iterative Learning Control was proposed for the pointwise tracking of a given continuous trajectory on a finite time interval. This requires the measurement of the tracking error along the whole time interval. However, due to insufficient measurement capability in many real control problems such as Rapid Thermal Processing (RTP) in wafer fab industry, it may happen that only the terminal output tracking error instead of the whole output trajectory tracking error is available. In this situation, applying ILC method is still possible as shown in this chapter. We refer to this special scheme as "Terminal Iterative Learning Control".

In this chapter, we concentrate on a typical terminal thickness control problem in RTP chemical vapor deposition (CVD) of wafer fab industry. Rapid thermal processing (RTP) systems are single-wafer, cold-wall chambers that utilize one or more radiant heat sources to rapidly heat up the semiconductor substrate at high temperatures for short times [212]. In semiconductor wafer fab industry, RTP has been used as a *versatile* single-wafer processing technique for various thermal processing applications. Single-wafer processing will be preferred over conventional batch equipment for many applications as the wafer size increases beyond 150-200 mm. Factors in favor of single-wafer processing include compatibility with multi-chamber cluster equipment for vacuum-integrated processing, improved fabrication cycle time, and enhanced fabrication process repeatability due to improved process control. One of the typical RTP systems is for the chemical vapor deposition (RTPCVD). The terminal deposition thickness (DT) for each run of RTPCVD is to be controlled within a given tolerance by adjusting the heating lamp power profile [61].

As the RTPCVD system executes a given task repeatedly, this repeatability can be utilized to improve the system control performance by the *iterative learning control* (ILC) method [17, 172]. In RTPCVD, however, the problem cannot be formulated in such a conventional way because the exact measurement of wafer temperature is almost impossible [189] even if a desired wafer temperature profile may be well planned which produces a desired terminal DT. The ultimate control objective in RTPCVD is to control the deposition thickness (DT) at the end of an RTP cycle. The control profile for the next operation cycle has to be updated using the terminal DT tracking error alone, i.e., the DT is measured only at the end of each RTP operation. It is possible to only use the terminal DT measurement while keep applying ILC-idea [61]. This is also called "point-to-point (PTP) control by iterative learning" which only consider the terminal output of the system instead of the *trajectory* over the whole time interval. The precise PTP control finds increasing applications in flexible manipulators [25], mobile robots [182] etc. Learning method has been applied to PTP control problems successfully in [162, 182]. The key technique is to represent the control function as a linear combination of a pre-determined piecewise continuous functional basis (orthogonal polynomials, splines etc.) and then to (iterate) update the coefficient vector based on the terminal output error measurement at the end of each run (system repetition). However, the existing results are only for continuous-time linear systems without robustness consideration.

Motivated by the control of RTPCVD system, this chapter presents a high-order Terminal ILC scheme for this new application of ILC method. By parameterizing the control profile with a piecewise continuous functional basis, the parameters are then updated by a high-order updating scheme. A convergence condition is established for a class of uncertain discrete-time time-varying linear systems. Simulation results are presented to demonstrate the effectiveness of the proposed high-order terminal iterative learning control scheme.

This chapter is organized as follows. An RTPCVD model is introduced in Sec. 7.3 which is linearized and discretized into a linear time-varying discrete-time system. In Sec. 7.2, a high-order terminal iterative learning control problem is formulated with a convergence analysis. Simulation results are presented in Sec. 7.4 to illustrate the effectiveness of the method proposed. Concluding remarks are given in Sec. 7.5.

7.2 Terminal Output Tracking by Iterative Learning

Consider the uncertain time-varying linear discrete-time system

$$\begin{cases} x_i(t+1) = A(t)x_i(t) + B(t)u_i(t) + w_i(t) \\ y_i(t) \quad\;\; = C(t)x_i(t) + v_i(t) \end{cases} \tag{7.1}$$

where $t = 0, 1, \cdots, N$; i is the system iteration number; state $x(t) \in R^n$; control function $u(t) \in R^m$; output $y(t) \in R^r$; $w_i(t), v_i(t)$ are model uncertainty and measurement disturbance respectively. Suppose the system (7.1) is completely controllable and observable. Assume that only $y_i(N)$ is measurable at the end of every run i. The control task is to update the control function $u_i(t)$ in an iterative manner such that $y_i(N)$ approaches to a given terminal output y^d as i increases.

As the tracking control task is only for the terminal output, conventional ILC updating law [17] can not be applied. Following the idea of **dynamic**

fitting [52], instead of solving a functional minimization problem, we parameterize the steering control $u_i(t)$ as

$$u_i(t) = \Phi(t)\Xi_i \tag{7.2}$$

where $\Xi_i = [\xi_{1_i}, \cdots, \xi_{p_i}]^T \in R^{p \times 1}$ is a parameter vector; $\Phi(t) = [\phi_1(t), \cdots, \phi_m(t)]^T \in R^{m \times p}$ is a properly chosen basis function matrix, $\phi_j(t) = [\varphi_{j1}(t), \cdots, \varphi_{jp}(t)]$, $j = 1, \cdots, m$. The task is hence converted into finding an iterative scheme based on the observation of the terminal output tracking error $e_T^{(i)} \triangleq y^d - y_i(N)$ such that the parameter Ξ_i converges as i increases and meanwhile $e_T^{(i)}$ converges to a prescribed bound centered at the origin.

Solving (7.1), one obtains

$$x_i(N) = Gx_i(0) + H\Xi_i + w_T^{(i)} \tag{7.3}$$

where $G \in R^{n \times n}$, $H \in R^{n \times p}$, $w_T^{(i)} \in R^{n \times 1}$.

$$G = \prod_{t=0}^{N-1} A(t); \tag{7.4}$$

$$H = \sum_{k=1}^{N} \prod_{j=1}^{k-1} A(N-j)B(N-k)\Phi(N-k); \tag{7.5}$$

$$w_T^{(i)} = \sum_{k=1}^{N} \prod_{j=1}^{k-1} A(N-j)w_i(N-k). \tag{7.6}$$

Thus the terminal output becomes

$$y_i(N) = CGx_i(0) + CH\Xi_i + Cw_T^{(i)} + v_i(N). \tag{7.7}$$

Investigating the terminal tracking error, one can easily get

$$e_T^{(i+1)} = e_T^{(i)} - CG\Delta x_{i+1}(0) - CH\Delta\Xi_{i+1}$$
$$-C\Delta w_T^{(i+1)} - \Delta v_{i+1}(N) \tag{7.8}$$

where $\Delta h_{i+1} \triangleq h_{i+1} - h_i$, $h \in \{x(0), w_T, v(N), \Xi\}$.

The high-order learning updating law is proposed as follows:

$$\Xi_{i+1} = \Xi_i + \Delta\Xi_{i+1} = \Xi_i + \sum_{k=1}^{M} L_k e_T^{(i-k+1)} \tag{7.9}$$

where M is the order of ILC and $L_k(k = 1, 2, \cdots, M)$ are learning gain matrices which are to be specified in applications. Substituting (7.9) into (7.8) yields

$$e_T^{(i+1)} = (I_r - CHL_1)e_T^{(i)} - CG\Delta x_{i+1}(0)$$

$$-CH\sum_{k=2}^{M} L_k e_T^{i-k+1} - C\Delta w_T^{(i+1)} - \Delta v_{i+1}(N) \tag{7.10}$$

where I_r is an $r \times r$ unit matrix.

To restrict our discussion, the following assumptions are imposed:

- A1). The initial state $x_i(0)$ at every iteration i is to be repositioned at the same starting point. However, small misalignment is allowed in such a way that

$$\|\Delta x_{i+1}(0)\| \leq \varepsilon_1$$

where ε_1 is a small positive number.

- A2). The disturbance and modeling error are bounded, i.e.,

$$\sup_{t \in \{0, \cdots, N\}} \|w_i(t)\| \leq b_w, \quad \sup_{t \in \{0, \cdots, N\}} \|v_i(t)\| \leq b_v, \forall i.$$

Furthermore, it is required that

$$\|\Delta w_T^{i+1}\| \leq \varepsilon_2, \quad \|\Delta v_{i+1}(N)\| \leq \varepsilon_3$$

where ε_2 and ε_3 are small positive numbers.

- A3). CH has a full row rank.

The key issue is the assumption A3). This related to the existence of the control and has been well discussed in [24]. As indicated in [24], a proper choice of $\Phi(t)$ is always possible if the system considered is controllable. Two examples were given to illustrate the choices of $\Phi(t)$ in [24] which are however not unique. Through a judicious choice of the basis functions of control $u(t)$, certain optimality constraints such as minimal energy, minimal norm etc. are to be satisfied to make the obtained control action be unique.

Remark 7.2.1. In PTP control problems, a control parameterization [24], or *control shaping*, is shown to be effective. When the task is repetitive, the idea of iterative learning [17] can be applied. It is interesting to note that even in the continuous path following control, a similar input shaping has been proposed [222].

Remark 7.2.2. Consider the case when system (7.1) is SISO LTI and the relative degree is d_r. We know that $CA^j B = 0$, $j = 0, 1, \cdots, d_r - 1$. Referring to (7.5),

$$CH = \sum_{k=1}^{N} CA^{k-1} B\Phi(N - k).$$

Hence, for a properly chosen Φ, CH can be made full rank if $d_r < N$.

We summarize the main result of this chapter in the following theorem.

Theorem 7.2.1 *Consider an uncertain system (7.1) under assumptions A1)-A3) with a given achievable terminal output y^d. By applying a proper control functional parameterization (7.2) and iterative learning updating law (7.9), through the repetitive operations, the terminal tracking error $e_T^{(i)}$ is bounded if*

$$\rho \triangleq \sum_{k=1}^{M} \rho_k < 1 \tag{7.11}$$

where $\rho_1 = \|I_r - CHL_1\|$, $\rho_k = \|CHL_k\|$ $k = 2, 3, \cdots, M$. Moreover, ε^*, the convergence bound of $e_T^{(i)}$ as $i \to \infty$, is a class-K function of $\varepsilon_1, \varepsilon_2$ and ε_3, which means $\varepsilon^* \to 0$ as $\varepsilon_1, \varepsilon_2$ and ε_3 approach to 0.

Now we proceed to prove Theorem 7.2.1.

Proof: Taking norm operation of (7.10) gives

$$\|e_T^{(i+1)}\| \leq \sum_{k=1}^{M} \rho_k \|e_T^{(i-k+1)}\| + \|CG\|\varepsilon_1$$
$$+ \|C\|\varepsilon_2 + \varepsilon_3. \tag{7.12}$$

The proof ends by applying Lemma 2.3.1 directly.

Remark 7.2.3. If the uncertainty $w(t)$ is repeatable from iteration to iteration, their effects on the ILC can be rejected according to Theorem 7.2.1. The neglected high order terms $w_1(t), w_2(t)$ in the linearized model (7.19) can be regarded as repetitive when a good initial control is used. Moreover, as shown in [57], it is also possible to track a slowly varying y^d by the method proposed in this chapter.

Remark 7.2.4. Consider the case when system (7.1) is LTI and the relative degree is d_r. We know that $CA^j B = 0, j = 0, 1, \cdots, d_r - 1$. Referring to (7.5),

$$CH = \sum_{k=1}^{N} CA^{k-1}B\Phi(N - k).$$

Hence, for a properly chosen Φ, CH can be made full rank if $d_r < N$. Usually, the number of sampling points is much larger than the relative degree. Thus, A4) can be satisfied easily.

Remark 7.2.5. Regarding the Φ selection, in general, only assumption A4) needs to be satisfied. In practice, to choose a suitable $\Phi(t)$, we need to have some knowledge about the system to be controlled. Φ can be chosen in terms of the particulars of the plant so as to further improve the control performance. A detailed example is presented in Sec. 7.4.

7.3 RTPCVD Model and Terminal Iterative Learning Control

A simplified model for RTPCVD of poly-Si, which includes the temperature dependence of deposition rate [94, 256], is considered as follows:

$$\begin{cases} \frac{dT_w}{dt} = [\sigma A_w E_w (T_{amb}^4 - T_w^4) + f E_w Q P]/M_w \\ \frac{dS}{dt} = k_0 \exp(-\frac{\gamma}{RT_w}) \end{cases} \tag{7.13}$$

where the meanings of variables and the relevant parameters are given in Table 7.1. $T_w(0)$ is known and $S(0) = 0\mu m$.

Based on the conventional Iterative Learning Control, the control objective is to find a lamp power profile $P(t)$ such that the controlled wafer temperature $T_w(t)$ follows the pre-planned one $T_w^d(t)$ as tight as possible in a given time interval $[0, NT_s]$, where T_s is the sampling period and N is the number of samples. According to [256], the $T_w^d(t)$ is pre-designed to satisfy the RTP requirements and especially to guarantee the final deposition thickness $S(T)$ $(T = NT_s)$. However, this control task is impractical because the *in-situ* measurement of wafer temperature $T_w(t)$ is an even more difficult problem. A practical way is to use the measurement of terminal DT only at the end of each iteration. In this case, the control task can be stated as follows:

Given the RTP cycle period T and a desired DT $S_d(T)$, with the repetitive operations of RTP process, iteratively update the control $P(t)$ based on the terminal DT tracking error $e_T \triangleq S_d(T) - S(T)$. To make the problem tractable, the control function $P(t)$ is parameterized by

$$P(t) = \Phi(t)\Xi = \sum_{j=1}^{p} \phi_j(t)\xi_j \tag{7.14}$$

where $\phi_j(t)(j = 1, \cdots, p)$ are known basis functions and the parameters $\xi_j(j = 1, \cdots, p)$ are to be determined through iterative learning. A simple updating law for Ξ is that

$$\Xi_i = \Xi_{i-1} + Le_T^{(i-1)} \tag{7.15}$$

where i is the iteration number and L is the learning gain matrix to be designed. A learning convergence condition for L is discussed in a general setting in Sec. 7.2.

It should be pointed out that a good choice of $\Phi(t)$ and the initial Ξ_0 is important to the above scheme. Although the actual RTPCVD process is complex, the simplified model (7.13) is still useful in choosing a good $\Phi(t)$, Ξ_0 and L. As $S(t)$ is a monotonically increasing function, one may approximate it by $\hat{S}(t)$ which can be taken as a quadratic polynomial or an exponential function with two unknowns. The two unknowns can be determined by the two known boundary conditions $S(T)$ and $T_w(0)$. Hence, with this obtained

$\hat{S}(t)$, one may estimate $T_w(t)$ and $P(t)$ by $\hat{T}_w(t)$ and $\hat{P}(t)$ respectively. These in turn can be used to determine $\Phi(t)$ and Ξ_0.

For example, set approximately $\hat{S}(t) = \alpha(e^{\beta t} - 1)$. One can get α, β by setting

$$\begin{cases} \alpha(e^{\beta T} - 1) = S_d(T), \\ \alpha\beta = \mathrm{d}\hat{S}(t)/\mathrm{d}t \mid_{t=0} = k_0 exp(-\gamma/[RT_w(0)]). \end{cases} \qquad (7.16)$$

With this approximation, one can immediately get

$$\hat{T}_w(t) = 1/(\alpha' + \beta't) \qquad (7.17)$$

where $\alpha' = -R[\ln(\alpha\beta/k_0)]/\gamma$ and $\beta' = -R\beta/\gamma$. By substituting $\hat{T}_w(t)$ into (7.13), the estimated control can be written in the following form:

$$\hat{P}(t) = c_0 + c_1/(\alpha' + \beta't)^2 + c_2/(\alpha' + \beta't)^4 \qquad (7.18)$$

where c_0, c_1, c_2 are known constants based on (7.13). Then a possible choice of $\Phi(t)$ is $[1, 1/(\alpha' + \beta't)^2, 1/(\alpha' + \beta't)^4]^T$.

The above simple derivations show how to choose $\Phi(t)$ and Ξ_0 based on a simplified RTPCVD model. However, the more important task is to determine the learning gain L. To obtain a suitable L, the results in previous section can be used. To do so, one may linearize (7.13) around an equilibrium point $(T_w^*(t), S^*(t), P^*(t))$ which can be obtained approximately by the procedures introduced in the above. A small variation around the equilibrium point is denoted by $(\Delta T_w(t), \Delta S(t), \Delta P(t))$. The linearized system is

$$\begin{cases} \frac{\mathrm{d}\Delta T_w}{\mathrm{d}t} = [-4\sigma A_w E_w (T_w^*)^3 \Delta T_w) + f E_w Q \Delta P]/M_w + w_1(t) \\ \frac{\mathrm{d}\Delta S}{\mathrm{d}t} = k_0 \frac{\gamma}{R} \exp(-\frac{\gamma}{RT_w^*}) \Delta T_w/(T_w^*)^2 + w_2(t) \end{cases} \qquad (7.19)$$

where $w_1(t), w_2(t)$ are the high order terms which can be taken as modeling uncertainties. Standard discretization of the linearized system (7.19) gives a time-varying SISO system.

Remark 7.3.1. The boundedness of trajectories during each run is guaranteed theoretically because only a finite time interval is concerned in every run. However, in practice, it may happen that the tracking errors in between sampling points are large while maintaining a good pointwise tracking. In such cases, properly choosing $\Xi^{(0)}$ and $\Phi(t)$ is essential. Fully utilizing the known information about the system may be helpful as shown in the simulation study presented in the following.

7.4 Simulation Studies

To simulate the actual situation, the following more practical RTPCVD model which considers the quartz window effect is used:

$$\begin{cases} \frac{dT_w}{dt} = [\sigma A_w E_w (T_q^4 - T_w^4) + f E_w Q P]/M_w \\ \frac{dT_q}{dt} = [QP + hA_q(T_{amb} - T_q)]/M_w \\ \frac{dS}{dt} = k_0 exp(-\frac{\gamma}{RT_w}) \end{cases} \qquad (7.20)$$

where the related parameters are listed in Table 7.1. However, the simplified model (7.13) is used for the ILC design. The initial states are $T_w(0) = T_q(0) = 300°K$, $S(0) = 0\mu m$. $T = 220s$. The desired final DT is 0.5 μm. The objective is to find the lamp power profile $P(t)$ by using the idea of iterative learning.

A well pre-designed wafer temperature profile can be used [61]. Such a wafer temperature profile will give a final DT (deposition thickness) of 0.5 μm. However, online measurement of $T_w(t)$ is required for the closed loop control. To overcome the difficulty in $T_w(t)$ measurement, we apply the idea of ILC and the terminal DT measurement only.

Table 7.1. Parameters for an RTPCVD system

symbol	value	meaning	unit
A_q		quartz window area	cm^2
A_w	400	wafer area	cm^2
E_w	0.8	wafer emissivity	unitless
f	0.5	lamp power absorbed by wafer	unitless
h		heat transfer coefficient for forced convection	$cal/cm^2/s/°K$
P	$\in [0,1]$	lamp power control factor	unitless
Q	1076	lamp power constant	cal/s
R	1.9872	gas constant	$cal/(gmol·°K)$
S		polysilicon deposition thickness	μm
T_w		wafer temperature	°K
T_q		quartz window temperature	°K
T_{amb}		ambient temperature	°K
M_w	1	wafer thermal mass	cal/°K
M_q	100	quartz window thermal mass	cal/°K
k_0	591000	pre-exponential constant of polysilicon deposition	$\mu m/s$
γ	39200	activation energy of polysilicon deposition	cal/gmol
σ	1.356×10^{-12}	Boltzmann constant	$cal/(s·cm^2·°K^2)$
		$hA_q \triangleq 1.84$ cal/s/°K	

The lamp power profile $P(t)$ is parameterized according to (7.18). Based on Table 7.1, $c_0 = -0.0082$; $c_1 = 2.8 \times 10^{-8}$; $c_2 = 1.01 \times 10^{-12}$; $\alpha' = 0.0033$; $\beta' = -1.202 \times 10^{-5}$. We know that $\Phi(t) = [1, 1/(\alpha' + \beta' t)^2, 1/(\alpha' + \beta' t)^4]$ and Ξ_0 can be chosen as $[c_0, c_1, c_2]'$.

Taking $P^{(0)}(t) = \Phi(t)\Xi_0$ as a nominal control and then linearizing the plant, we can get a discrete-time model similar to (7.1). This enables us to choose the learning gains L_k. For convenience, we use $L_1 = K\Xi_0$ where

K is a constant. Based on the convergence analysis in Sec. 7.2, we know $K > 0$. Moreover, if the plant model (7.20) is close to the simplified one (7.13), $K = 1$ will give a good convergent learning process. However, (7.20) is used in the simulation while (7.13) is employed in the analysis and design of learning controller. Furthermore, the convergence condition for learning gains is based on a linearized model. Due to these approximations, K should be changed from 1. Fig. 7.1 summarizes the convergence comparison of terminal DT tracking by the first-order ($M = 1$) iterative learning schemes with different K's. A good convergence result can be observed for $K = 10$. Larger K may result in unacceptable oscillations while smaller K slows down the convergence. Fig. 7.1 also illustrates that the ILC convergence is robust to model uncertainty and disturbance.

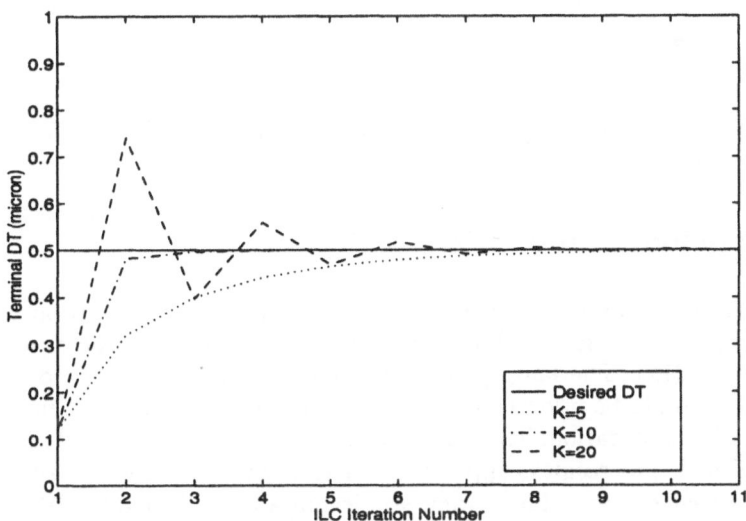

Fig. 7.1. Convergence comparison of terminal DT tracking by the first-order iterative learning schemes.

In practice, one may use some tuning method to schedule different values of K in different iterations which is similar to the PID tuning in the iteration number i-axis. As the first-order ILC updating law is in fact a pure integral controller along the i-axis, higher order ILC scheme may be used which results in PI or PID type controller in the i-axis. Intuitively, better results can be expected. To illustrate this, a second order scheme is used and L_2 is simply set to $10\% L_1$. Improved results can be clearly observed in Fig. 7.2 for both $K = 20$ and $K = 5$. It is interesting to note that, when $K = 20$, the oscillation has been improved a lot by using a high-order updating law.

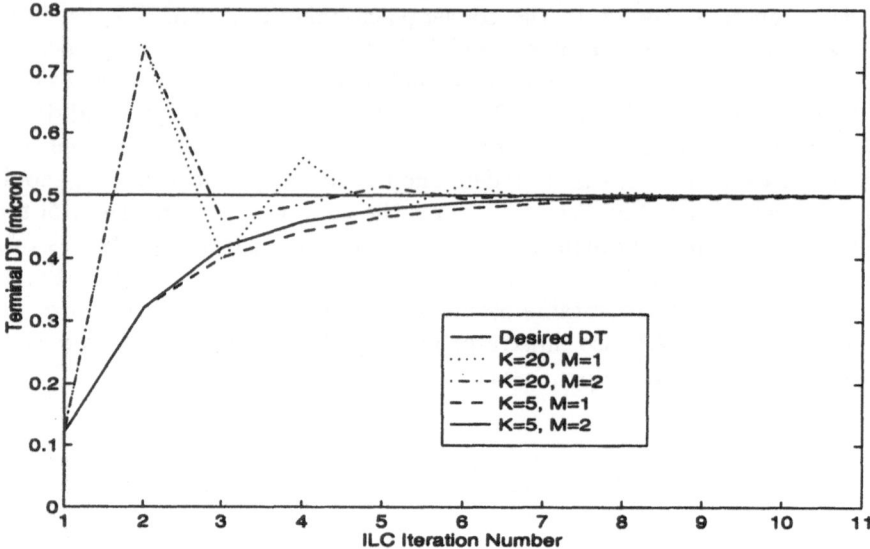

Fig. 7.2. Convergence comparison of terminal DT tracking by high-order iterative learning schemes.

7.5 Concluding Remarks

A high-order terminal iterative learning control algorithm is proposed for the terminal deposition thickness control for rapid thermal process chemical vapor deposition. The lamp power control profile is parameterized and the parameters are to be updated by using the terminal DT measurement only at the end of each iteration. The convergence condition of the control algorithm is established. The result is applied to a class of uncertain discrete-time time-varying linear systems. Simulation studies on a simplified RTPCVD model show that the desired DT can be achieved by the proposed terminal output learning scheme in a few iterations. It is also illustrated that a high order scheme can provide better convergence performance.

8. Designing Iterative Learning Controllers Via Noncausal Filtering

8.1 Introduction

MOST OF THE EXISTING WORK focuses on the *analysis* issue of ILC schemes while the obtained convergence condition is clearly not enough for actual ILC applications. e.g., [46, 251, 140]. Therefore, in recent years, increasing efforts have been made on the *design* issue of ILC. A recent survey on ILC *design* issue [155] documented various practically tested design schemes mainly for robotics applications where detailed descriptions can be found in [84]. Other efforts in ILC design include using modern techniques like LMI (linear matrix inequality) [73], H^∞ [12], frequency domain LFT (linear fractional transformation) [171], local ARMA (auto-regressive moving average) model approximation [192]. However, to draw the attention from the industry, the above mentioned design techniques are still not sufficient as compared to the successful and wide application of PID (proportional-integral-derivative) controllers in industries.

According to a survey [255] on the state of process control systems, more than 90% of the control loops were of the PID type. It was also indicated [26] that a typical chapter mill in Canada has more than 2000 control loops and in which 97% use PI control. Therefore, the industrialist had concentrated on PI/PID controllers and had already developed *one-button type* relay auto-tuning techniques for fast, reliable PI/PID control yet with satisfactory performance [148, 233]. In view of the above fact, the ILC design problem should also be attacked in a similarly rudimentary way as the PI/PID controllers. Other advanced techniques could be accomplished as a value-added block.

It is interesting to note that the noncausal filtering [70, 84, 155], especially zero phase filtering [187, 33] technique may provide a way for such a rudimentary design task as discussed above. This chapter puts efforts in this direction aiming at providing a PID-autotuning alike ILC design procedure. Convergence analysis has been given in detail and the design steps are clearly given explicitly under assumptions close to actual application situations. Some practical considerations in the parameter tuning are also outlined. A limit on the ILC convergence rate has been obtained for the first time.

In this chapter, an l_1-norm on a vector space $\mathcal{V} \subseteq R^{N \times (N_p+1)}$ is defined as follows: for every $v \in \mathcal{V}$, $v = \{v_{j,k} \mid j = 1, \cdots, N; k = 0, 1, \cdots, N_p\}$,

$$\|v\|_1 \triangleq \sum_{k=0}^{N_p} \|v_k\|$$

where $\|\cdot\|$ is a usual vector norm, e.g.,

$$\|v_k\| \triangleq \{\sum_{j=1}^{N} (v_{j,k})^2\}^{\frac{1}{2}}.$$

In this chapter, we assume $\mathcal{V}$ contains a series of function vectors on sampling instants. These functions in continuous time domain are supposed to be $\mathcal{C}^2$, i.e., twice continuous and differentiable for all t with some exceptions in a finite set $\mathcal{D}$. In actual computer controlled systems, most signals belong to $\mathcal{V}$ where N_p is the number of samples and N the dimension of the signal vector.

8.2 Noncausal FIR Filter and Its Properties

Filtering is a convolution between the input signal and the filter's impulse response of the filter. A causal filter is defined as one for which there is no output prior to the time of non-zero input. Consider a discrete filter with impulse response denoted by h_k where k is an integer. A causal FIR (Finite Impulse Response) filter requires that $h_k = 0$ when $k < 0$. For any input signal $x(k)$, the filter output $y(k)$ can be expressed by

$$y(k) = \sum_{s=0}^{M} h_s x(k-s) \triangleq h_M * x(k) \tag{8.1}$$

where M is the length of the FIR filter and '*' denotes the discrete convolution. Therefore, 'h_M*' is a *mapping operator* between input and output signals. Using the notation of $\mathcal{Z}$-transform, (8.1) can be written as

$$Y(z) = H(z) \cdot X(z)$$

where $Y(z) \triangleq \mathcal{Z}(y(k)); X(z) \triangleq \mathcal{Z}(x(k)); H(z) \triangleq \mathcal{Z}(h_k)$ and $H(z) \triangleq \sum_{s=0}^{M} h_s z^s$ denoting the $\mathcal{Z}$-transfer function of the FIR filter. The above discrete FIR is also called an "MA" (moving average) filter. It is clear that FIR filters are always stable since their impulse responses have finite energy.

For noncausal FIR filters, h_k can be nonzero when $k < 0$. In this chapter, only a class of simple noncausal FIR filters is considered. Given the causal FIR filter $H(z)$, the corresponding zero phase (noncausal) filter is $H'(z) = H(z) \cdot H(z^{-1})$, i.e.,

$$H'(z) = \sum_{k=-M}^{M} h'_k z^{-k} = \sum_{k=0}^{M} h_k z^k \cdot \sum_{k=0}^{M} h_k z^{-k}. \tag{8.2}$$

It is easy to show that

$$\begin{cases} h'_0 = \sum_{k=0}^{M} h_k^2; \\ h'_k = h'_{-k}; \\ h'_k = \sum_{\substack{i-j \le k \\ i=0,1,\cdots,M \\ j=0,1,\cdots,M}} h_i h_j. \end{cases} \tag{8.3}$$

The above zerophase FIR filter operator h'_M* is actually a symmetric moving average. Usually, the coefficients are normalized to 1, i.e.,

$$2\sum_{k=1}^{M} h'_k + h'_0 = 1. \tag{8.4}$$

We will show an interesting property in the following lemma.

Lemma 8.2.1 *Consider the zero phase filter h'_M* of length $2M+1$ with its coefficients normalized as in (8.4). For any integer k, we have*

$$\sum_{j=k-M}^{k+M} h'_{j-k} \cdot (j-k) = 0. \tag{8.5}$$

Proof. The proof is straightforward using (8.4) and the symmetry of coefficients h'_k.

In what follows, we will show an important property of the zero phase filter operator discussed above which is crucial in the convergence analysis of the proposed learning controller.

Consider a continuous-time signal $g(t), t \in [0,T]$. When sampled at the sampling period of T_s, it is assumed that $g(kT_s) \in \mathcal{V}, N_p T_s = T$. When applying the above zero phase filter h'_M* on $g(kT_s)$, a contraction mapping can be constructed as shown in the following lemma.

Lemma 8.2.2 *Consider the zero phase filter h'_M* of length $2M+1$ with its coefficients normalized as in (8.4). For any $g(kT_s) \in \mathcal{V}, N_p T_s = T$, there exists a positive real number γ, which is a function of M, such that operator $(1 - \gamma h'_M*)$ is a contraction mapping on $g(\cdot)$, i.e.,*

$$\|(1 - \gamma h'_M*)^i g(\cdot)\|_1 \le \rho^i \|g(\cdot)\|_1 \to 0 \text{ as } i \to \infty \tag{8.6}$$

where $\rho < 1$.

Proof. First, we assume $\mathcal{D} = \emptyset$. From the fact that $g(t)$ is $\mathcal{C}^2$, using the second order Taylor expansion of $g(t)$ around kT_s gives

$$g(t) = g(t)\mid_{kT_s} + \frac{dg}{dt}\mid_{kT_s} (t - kT_s) + \mathcal{O}((t - kT_s)^2). \tag{8.7}$$

For every element g_j of g, applying the zero phase filter operator h'_M* yields

$$h'_M * g_j = g_j + \frac{dg_j}{dt} \sum_{j=k-M}^{k+M} h'_j(j-k)T_s + \mathcal{O}((MT_s)^2). \tag{8.8}$$

Applying Lemma 8.2.1, one obtains from (8.8) that

$$h'_M * g_j = g_j + \mathcal{O}((MT_s)^2).$$ (8.9)

Therefore, for any $\gamma > 0$,

$$(1 - \gamma h'_M *)g = (1 - \gamma)g + \gamma\mathcal{O}((MT_s)^2).$$ (8.10)

Estimating the l_1-norm of $(1 - \gamma h'_M *)g$,

$$\|(1 - \gamma h'_M *)g\|_1 \leq (1 - \gamma)\|g\|_1 + \gamma N_p \mathcal{O}((MT_s)^2).$$ (8.11)

Equation (8.11) can also be written as

$$\frac{\|(I_N - \gamma h'_M *)g\|_1}{\|g\|_1} \leq 1 - \gamma + \frac{\gamma N_p \mathcal{O}((MT_s)^2)}{\|g\|_1} \triangleq \rho.$$ (8.12)

Clearly, if $\frac{\gamma N_p \mathcal{O}((MT_s)^2)}{\|g\|_1}$ is small enough, there exists a $\gamma \in (0, 2)$ such that $|\rho| < 1$. Therefore, there exists a $\gamma(M) \in (0, 2)$ such that the operator $(I - \gamma h'_M *)$ is a contraction mapping on $g(\cdot)$ which verifies (8.6).

When $\mathcal{D} \neq \emptyset$, g is not C^2 on $\mathcal{D}$. In (8.12), the following term should be added

$$((2M + 1)\max_j \|g_j\|\|\mathcal{D}\|T_s)/\|g\|_1$$

which is normally quite small ($\ll 1$) when T_s is sufficiently small. Therefore, this does not alter the above conclusion due to that fact that $\mathcal{D}$ is finite.

Remark 8.2.1. It should be pointed out that the boundary effects in the filtering operator $(1 - \gamma h'_M *)$ should be considered in practical applications. Simply leaving the M points at both end of the signal array may be acceptable when no other prior information on the signal is available. In practice, to handle the boundary effects, the method used in filtfilt in MATLAB™ Signal Processing Toolbox can be applied.

8.3 Noncausal FIR Filtering Based Iterative Learning Controller

ILC, originally proposed as an open-loop control [17], has been considered as a feedforward control in addition to an existing feedback controller. While the most of the 'analysis' results were obtained in continuous-time domain, e.g., [100, 132, 142, 145, 120], discrete-time analysis, e.g., [58, 63], is more important since the realization of ILC algorithm is memory based. The feedforward-feedback configuration of ILC algorithms has already been a standard in either 'analysis' or 'design' [155] work on ILC. A block-diagram is shown in Fig. 8.1 where **FBC** stands for "feedback controller" and y_d is the given desired output trajectory to be tracked. After the i-th iteration (repetitive operation), the feedforward control signal u_{ff}^i and the feedback control signal u_{fb}^i are to be stored in the memory bank for constructing the feedforward control signal at the next iteration, i.e., u_{ff}^{i+1}. The stored feedback control signal u_{fb}^i are to be filtered by a noncausal filter operator $h_M'*$ multiplied by a learning gain γ.

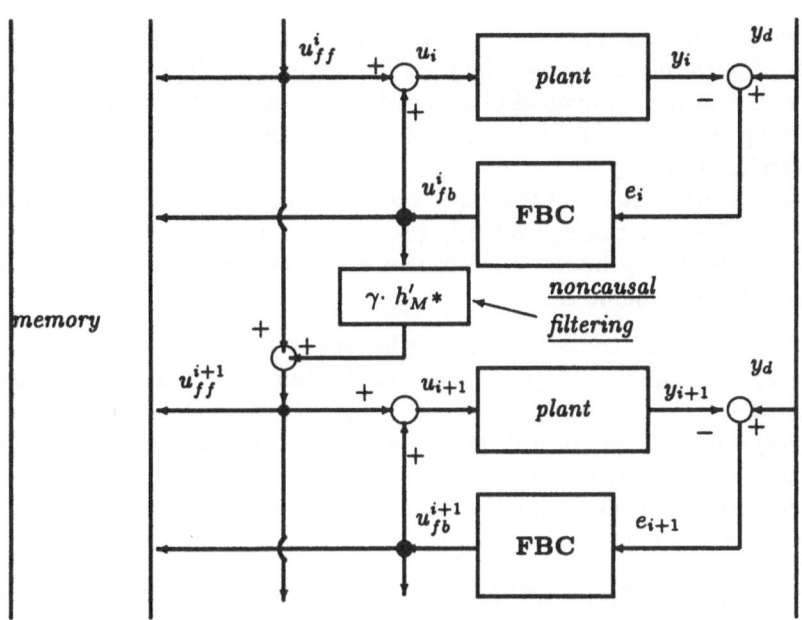

Fig. 8.1. Block diagram of noncausal filter based iterative learning control

The learning updating law is hence written as

$$u_{ff}^{i+1}(k) = u_{ff}^i(k) + \gamma h_M' * u_{fb}^i(k) \tag{8.13}$$

while the overall control is simply that

$$u_{i+1}(k) = u_{ff}^{i+1}(k) + u_{fb}^{i+1}(k). \tag{8.14}$$

A complete block diagram is shown in Fig. 8.1 where two parameters - γ the learning gain and M the length index of the noncausal filter (8.2), are to be designed and specified.

The simplest form of (8.13) is that [33]

$$u_{ff}^{i+1}(k) = u_{ff}^i(k) + \frac{\gamma}{2M+1} \sum_{j=-M}^{M} u_{fb}^i(k+j) \tag{8.15}$$

where h_M' is simply an algebraic averager.

In this chapter, the detailed form of the dynamic system shown as "*plant*" in Fig. 8.1 will not be specified. This is not necessary as argued in [179] from the advocated **Principle of Self Support** (PSS) because the **stored** control signals are in essence the *"plant"* itself for the specific control task. Control problem is with *self-circularity* just as a twofold uroboros [179, page 23]. Taking a rigid robot as an example, the system dynamics derived using Lagrange method can be written in a parameterized form as

$$\tau = \Psi(q, \dot{q}, \ddot{q})p \tag{8.16}$$

where τ is the control torque vector and p is the parametric vector containing only physical parameters of the robotic manipulator. It is an important concept that in (8.16), instead of always being thought of τ as a *cause* and the tracking error the *effect* or *consequence*, τ in fact fully represents the system dynamics [179].

Therefore, instead of making some assumptions on the system (class) to be controlled, we can equivalently preassume some properties on the control signals which may be more direct and practical. For practical systems, it is reasonable to assume that

- (A1) The plant to be controlled is internally stable;
- (A2) The number of input is equal to the number of output (measurement variable);
- (A3) The desired control input $u_d(t)$ exists uniquely for a given desired output trajectory $y_d(t)$;
- (A4) The system dynamics $u_i(t)$ can be decomposed into a repeatable part $u_d^R(t)$ and a nonrepeatable part $u_i^{NR}(t)$, i.e.,

$$u_i(t) \overset{\triangle}{=} u_d^R(t) + u_i^{NR}(t) \tag{8.17}$$

where $u_i^{NR}(t)$ is bounded by

$$h_M' * u_i^{NR} \leq \varepsilon^*, \quad \forall i. \tag{8.18}$$

Remark 8.3.1. If $u_i^{NR}(t)$ is a random variable with zero mean, $h_M'*$ takes the special form of (8.15), in this case, (8.18) becomes

$$h_M' * u_i^{NR} \approx \varepsilon^*, \quad \forall i. \tag{8.19}$$

where $\varepsilon^* = 0$. Therefore, in actual applications, ε^* is normally small.

8.4 A Convergence Analysis

The convergence of the proposed learning controller is in the sense that u_{ff}^i approaches to u_d^R as i increases. This is summarized in the following theorem.

Theorem 8.4.1 *A system satisfying assumptions (A1)-(A4) is controlled by a suitable feedback controller as shown in Fig. 8.1 which performs a given task repeatedly. When applying noncausal filtering based ILC scheme (8.13), there exists a $\gamma(M) \in (0,2)$ such that the learning process is convergent, i.e.,*

$$\lim_{i\to\infty} \|u^i_{ff}\|_1 \to \|u^R_d\|_1 + \varepsilon_0$$

where ε_0 is a class-$\mathcal{K}$ function of ε^*.

Theorem 8.4.1 implies that the iterative learning controller is essentially applied for the *extraction* of the repeatable part of the system dynamics u^R_d while the nonrepeatable part is taken care of by the feedback controller. When the noncausal filter based ILC scheme (8.13), shown in Fig. 8.1, is used, there always exists a choice of learning gain γ which is a function of the length index of the filter such that the learning process converges.

With the explored properties of noncausal filter in Sec. 8.2, the proof of Theorem 8.4.1 becomes an easy task.

Proof. From (8.14), the feedback signal at the i-th iteration can be written as

$$\begin{aligned} u^i_{fb} &= u_i - u^i_{ff} \\ &= u^R_d + u^{NR}_i - u^i_{ff}. \end{aligned} \tag{8.20}$$

Applying the filtering operator $h'_M *$ on both sides of (8.20) gives

$$h'_M * u^i_{fb} \le h'_M * u^R_d - h'_M * u^i_{ff} + \varepsilon^*. \tag{8.21}$$

Substituting (8.21) into the learning updating law (8.13), one obtains

$$u^{i+1}_{ff} \le (I - \gamma h'_M *)u^i_{ff} + \gamma(h'_M * u^R_d + \varepsilon^*). \tag{8.22}$$

Iterating i in (8.22),

$$\begin{aligned} u^i_{ff} &\le (I - \gamma h'_M *)^i u^0_{ff} + [I - (I - \gamma h'_M *)^{i+1}]u^R_d \\ &\quad + \frac{I - (I - \gamma h'_M *)^{i+1}}{h'_M *}\varepsilon^*. \end{aligned} \tag{8.23}$$

From Lemma 8.2.2, it has been shown that

$$(I - \gamma h'_M *)^i \to 0 \text{ when } i \to \infty,$$

when $u^R_d \in \mathcal{V}$ and $\gamma(M) \in (0,2)$.

Therefore, (8.23) becomes

$$\lim_{i\to\infty} \|u^i_{ff}\|_1 \le \|u^R_d\|_1 + \|\varepsilon^*\|_1. \tag{8.24}$$

When ε^* tends to 0,

$$\lim_{i\to\infty} u^i_{ff} = u^R_d. \tag{8.25}$$

Remark 8.4.1. From the above proof, it can be seen that the learning convergence is independent of u^0_{ff} which means that the initial feedforward control can be chosen arbitrarily. However, u^0_{ff} is practically set to 0 because no prior knowledge on determining u^0_{ff} is available.

Remark 8.4.2. The basic idea of the noncausal filtering based ILC is that the nonrepeatable part of the feedback signal is to be filtered while the remaining repeatable part is to be stored in the memory bank. By means of iterative learning, when approaching to the repeatable part of the (inverse) system dynamics, u_{ff}^i tends to relieve the burden of the feedback controller. This implies, critical or optimal design of a feedback controller may not be necessary.

8.5 ILC Design Method

Up to now, it has been shown that the design task of ILC can be reduced to the tuning of two parameters: length index M of the filter h'_M* and the learning gain γ. Meanwhile, whilst the simplest filter (8.15) is applicable, there is still a flexibility in choosing the shape of the filter h'_M*. In frequency domain, h'_M* can be shaped using the analog prototype low pass filter design and then converted to FIR discrete filter in the sense of weighted summation. Although this frequency domain approach may be more practical, a time domain direct design is more convenient with less efforts. In this chapter, h'_M* is given directly in discrete time domain with its normalized coefficients given by $h'_k (k = -M, \cdots, M)$.

Two frequently used noncausal filters are considered as examples for illustrating the design consideration. The time domain responses are shown in Fig. 8.2 where the rectangle one in thicker line, denoted by $h'_{M_1}(t)$, is the simplest algebraic averager described in (8.15). The triangle one is a *weighted* algebraic averager denoted by $h'_{M_2}(t)$.

In Fig. 8.2, $h'_{M_1}(t)$ and $h'_{M_2}(t)$ are normalized according to (8.4). With the specified $h'_M(t)$ as shown in Fig. 8.2, its frequency response can be computed. The Fourier transform of $h'_M(t)$ is denoted by $H'(\omega)$. For $h'_{M_1}(t)$ and $h'_{M_2}(t)$, one obtains

$$H'_{M_1}(\omega) = \frac{\sin(\omega M T_s)}{\omega M T_s}; \tag{8.26}$$

$$H'_{M_2}(\omega) = \frac{2[1 - \cos(\omega M T_s)]}{(\omega M T_s)^2}. \tag{8.27}$$

We consider the discrete-time case, i.e., $h'(-M), h'(-M+1), \cdots, h'(M-1), h'(M)$. The sampling period is T_s. The derivations of (8.26) and (8.27) are given as follows:

- *Derivation of (8.26).*
  The discrete time Fourier transformation for $h'_{M_1}(k), k = -M, \cdots, M$ can be expressed as

$$H'_{M_1}(j\omega) = \sum_{k=-M}^{M} h'_{M_1}(k) e^{-j\omega k T_s}$$

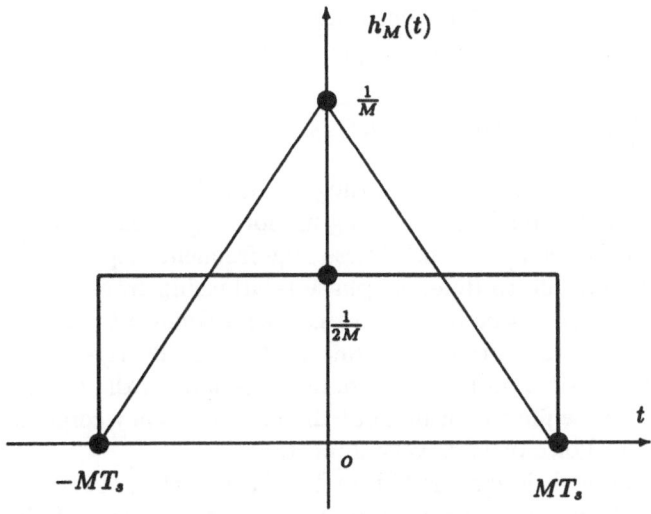

Fig. 8.2. Impulse responses of two simple zero phase filters

$$= \frac{1}{2MT_s} \sum_{k=-M}^{M} [e^{-j\omega kT_s}]T_s$$

$$\approx \frac{1}{2MT_s} \int_{-MT_s}^{MT_s} e^{-j\omega t} dt$$

$$= \frac{1}{2MT_s} \frac{-1}{j\omega} [e^{-j\omega MT_s} - e^{j\omega MT_s})$$

$$= \frac{\sin(\omega MT_s)}{\omega MT_s} = H'_{M_1}(\omega). \tag{8.28}$$

- *Derivation of (8.27).*
  Similar to the derivation of (8.26), a discrete time Fourier transformation can be actually obtained by a continuous time counterpart approximately.

$$H'_{M_2}(j\omega) \approx \frac{1}{T_s} \{ \int_{-MT_s}^{0} \frac{1/M}{MT_s}(t + MT_s)e^{-j\omega t} dt$$

$$+ \int_{0}^{MT_s} \frac{-1/M}{MT_s}(t - MT_s)e^{-j\omega t} dt \}$$

$$= \frac{1}{T_s} \int_{0}^{MT_s} \frac{1/M}{MT_s}(-t + MT_s)[e^{-j\omega t} + e^{j\omega t}] dt$$

$$= \frac{2}{M^2 T_s^2} \frac{1}{\omega} \int_{0}^{MT_s} (-t + MT_s) d\sin \omega t$$

$$= \frac{2}{M^2 T_s^2} \frac{1}{\omega} \int_{0}^{MT_s} \sin \omega t dt$$

$$= \frac{2(1 - \cos \omega MT_s)}{(\omega MT_s)^2} = H'_{M_2}(\omega). \tag{8.29}$$

8.5.1 Design Method for M

The basic consideration in choosing an M is that it should be neither too large nor too small. Small M will bring in more high frequency signal components stored in the memory bank. These high frequency signal components may be accumulated due to different phase relationship from iteration to iteration. This is the *key* reason of the divergence for some ILC schemes which may be convergent at the initial few iterations, become divergent as ILC continuously runs. A large M, on the other hand, deteriorates the signal's low frequency components while smoothing out the high frequency components. Therefore, a suitable choice of M is very important.

A practical design procedure can be like this. As at the first iteration only feedback controller is commissioned, at the end of the first iteration, performing discrete Fourier transform (DFT) of the feedback signal $u^0_{fb}(t)$ gives the spectrum information of the feedback signal. Therefore, the cut-off frequency ω_c of the filter can be obtained. ω_c should be chosen a little higher than the frequency corresponding to the magnitude peak in the amplitude-frequency plot of $u^0_{fb}(t)$. Using ω_c and setting

$$H'(\omega_c) = \frac{\sqrt{2}}{2} \tag{8.30}$$

will give an estimate of M for a given T_s.

For example, for $H'_{M_1}(\omega)$,

$$H'_{M_1}(\omega_c) = \frac{\sin(\omega_c MT_s)}{\omega_c MT_s} = \frac{\sqrt{2}}{2}$$

and one gets

$$\omega_c MT_s \approx 1.392 \tag{8.31}$$

by using `solve('sin(x)/x =sqrt(2)/2')` of MATLAB Symbolic Math Toolbox. Therefore,

$$M \approx \frac{1.392}{\omega_c T_s} \tag{8.32}$$

Similarly, from (8.27),

$$\omega_c MT_s \approx 2.004 \tag{8.33}$$

and

$$M \approx \frac{2.004}{\omega_c T_s} \tag{8.34}$$

by using `solve('2*(1-cos(x))/x^2 = sqrt(2)/2')`.

M could be enlarged a little bit in practice. However, note the fact that the learning gain γ is a function of M and that the large M will reduce the upper limit of $\gamma(M)$ as shown in (8.12). This means that γ has to be chosen in a very small interval, e.g., $(0, 0.1]$, instead of $(0, 2)$.

Remark 8.5.1. According to Shannon's sampling theorem, approximately, $\frac{2\pi}{\omega_c T_s}$ should larger than 2. Therefore, in any case, it can be concluded that M should be greater than $1.392/\pi$. This implies that when $M = 0$ and ω_c is finite, the ILC scheme

$$u_{ff}^{i+1}(k) = u_{ff}^i(k) + \gamma u_{fb}^i(k) \tag{8.35}$$

may not actually work well in practice.

8.5.2 Design Method for γ

How to design a suitable γ depends on what kind of knowledge is assumed about the plant to be controlled. How to get a reasonable estimate of the upper limit of γ with **less** effort is an open problem. Once this problem is solved, ILC design will be really *PID-tuning alike*. However, during practice, one can always start with a smaller, conservative γ via which the learning process converges, then fine tune γ which is to be discussed in Sec. 8.5.4. In general, it is easy to make ILC work.

In most engineering practice, it is quite common that the Nyquist curve information about the system is available. Since the system considered is controlled by a feedback controller, the closed-loop transfer function $G_c(j\omega)$ is assumed to be available. In this section, linear system notion is used because within certain range of frequencies concerned, from the Bode plot, it is always possible to approximate the practical system by a linear system.

Full knowledge of $G_c(j\omega)$, $\omega \leq \omega_c$ may be sometimes impractical. Therefore, it is assumed that at least $G_c(j\omega_c)$ is available. In what follows, it will be shown that $G_c(j\omega_c)$ can be used to design a reasonable γ.

Denote $G_b(j\omega)$ and $G_o(j\omega)$ the transfer functions of the feedback controller and the open-loop plant respectively. Then

$$G_c(j\omega) = \frac{G_b(j\omega)G_o(j\omega)}{1 + G_b(j\omega)G_o(j\omega)}. \tag{8.36}$$

Writing the ILC updating law (8.13) in the frequency domain gives

$$U_{ff}^{i+1}(j\omega) = U_{ff}^i(j\omega) + \gamma H'(\omega) U_{fb}^i(j\omega) \tag{8.37}$$

where U_* denotes the frequency domain counterpart of u_*. In the following, $(j\omega)$ is omitted for brevity. Since $U_{fb}(j\omega)$ can be written by

$$U_{fb} = \frac{G_b}{1 + G_b G_o} Y_d - \frac{G_b G_o}{1 + G_b G_o} U_{ff} \tag{8.38}$$

where Y_d is the frequency domain transformation of $y_d(t)$, (8.37) becomes

$$U_{ff}^{i+1} = (1 - \gamma H' G_c) U_{ff}^{i+1} + \gamma H' \frac{G_c}{G_o} Y_d. \tag{8.39}$$

Iterating (8.39),

$$U_{ff}^{i+1} = (1 - \gamma H' G_c)^i U_{ff}^0 + [1 - (1 - \gamma H' G_c)^{i+1}] \frac{1}{G_o} Y_d. \tag{8.40}$$

Clearly, the convergence condition is that

$$| 1 - \gamma H'(\omega) G_c(j\omega) | < 1, \quad \forall \omega. \tag{8.41}$$

The converged value is given by

$$U_{ff}^\infty = \frac{1}{G_o} Y_d \tag{8.42}$$

which requires an inverse of G_o.

Condition (8.41) may not be applicable for all ω. Therefore, only ω less than a frequency of interest, say, ω_c is considered. Designing γ is then to solve

$$| 1 - \gamma H'(\omega) G_c(j\omega) | < 1, \quad \forall \omega \leq \omega_c. \tag{8.43}$$

Since $H'(\omega)$ is known as shown, for example, in (8.26), γ can be obtained from the knowledge of $G_c(j\omega)$. A plot of $\gamma(\omega)$ is available from (8.43). This plot is useful in selecting a suitable γ when different frequencies of interest are to be considered. Usually, with the only information of $G_c(\omega_c)$, one can still determine a suitable γ.

In any case, it is possible to tune γ to make the ILC convergence as fast as possible. However, as shown in Sec. 8.5.3, the ILC convergence rate has its limit governed by the closed loop system dynamics alone.

8.5.3 A Limit of ILC Convergence Rate

Let $G_c(j\omega) = A(\omega) \exp^{i\theta(\omega)}$. Convergence condition (8.41) can be written as

$$| 1 - \gamma H'(\omega) A(\omega) \exp^{i\theta(\omega)} |$$
$$= \{1 - 2\gamma H'(\omega) A(\omega) \cos \theta(\omega) + H'^2(\omega) A^2(\omega) \gamma^2\}^{\frac{1}{2}}.$$

Suppose the above is bounded by ρ^* and $\rho^* \in (0,1)$. Then,

$$1 - 2\gamma H'(\omega) A(\omega) \cos \theta(\omega) + H'^2(\omega) A^2(\omega) \gamma^2 \leq (\rho^*)^2. \tag{8.44}$$

To have a real solution of γ, one requires that

$$4[H'(\omega) A(\omega) \cos \theta(\omega)]^2 - 4[1 - (\rho^*)^2][H'(\omega) A(\omega)]^2 \geq 0 \tag{8.45}$$

which gives

$$(\rho^*)^2 \geq \sin^2 \theta(\omega). \tag{8.46}$$

The above inequality implies that, for a given ω of interest, the ILC convergence rate cannot be faster than the limit characterized by ρ^*. This limit is independent of learning filter operator. The only way to achieve a faster ILC convergence process is to well design the feedback controller $G_b(j\omega)$ to reduce the phase shift (phase lag) of the closed-loop system. This is an interesting conclusion that explains many phenomenon appeared in practice.

8.5.4 Heuristic Design Schemes

The following heuristic ideas may be helpful in tuning ILC parameters. When applicable, some rule-based methods could be used.

- Re-evaluate M at the end of every iteration. This will not cost a lot but can keep a tight monitoring of possible variations of the system dynamics and the uncertainty/disturbance.
- When ILC starts with a smaller γ, increase γ if the tracking error keep decreasing and decrease γ if the tracking error keep increasing.
- Use a cautious M at the beginning of ILC process. Decrease M when ILC converges to a stage with little improvement. In this case, smaller M leaves more high frequency components of the feedback control signal in the memory bank. This in turn may further improve the convergence performance.
- The feedback controller's gain can be increased as ILC converges. In this case, the repeatable parts have been learned and the system under control is actually *linearized*.

8.6 Conclusions

It is advocated in this chapter that a PID-autotuning alike ILC design procedure is important for ILC to be acceptable by industry. This chapter has made some efforts in this direction by using the idea of noncausal FIR filtering techniques. The proposed learning scheme is very simple, straightforward and yet not model based. Convergence analysis has been given in detail for a class of zero phase filters. ILC design steps are given explicitly which is close to actual application situations. It is shown that the design task of ILC can be reduced to tuning two parameters: length of the filter and the learning gain. Some practical considerations in the parameter tuning are also outlined. Also, a limit on the ILC convergence rate has been established.

9. Practical Iterative Learning Control Using Weighted Local Symmetrical Double-Integral

9.1 Introduction

THE DESIGN ISSUE of iterative learning controller has been attacked in Chapter 8 using the *noncausal filtering* technique. It provides a systematic design procedure for ILC which acts as a learning feedforward control (LFFC). A successful application example can be found in [234] where *less* modeling effort is required with just a simple relay feedback experiment.

In this chapter, we presents a parallel result of Chapter 8 in continuous-time domain in terms of *local symmetrical integral*. A weighted local symmetrical double-integral (WLSI^2) of feedback control signal of the previous iteration is applied. The ILC updating law takes a simple form also with only two design parameters: the learning gain and the range of local double-integration. Convergence analysis is presented together with a design procedure. Some practical considerations in the parameter tuning are also outlined.

9.2 WLSI^2-type ILC

A block-diagram of the proposed practical iterative learning control scheme using the weighted local symmetrical double-integral is shown in Fig. 9.1 where **FBC** stands for "feedback controller" and y_d is the given desired output trajectory to be tracked. After the i-th iteration (repetitive operation), the feedforward control signal u_{ff}^i and the feedback control signal u_{fb}^i are to be stored in the memory bank for constructing the feedforward control signal at the next iteration, i.e., u_{ff}^{i+1}. The stored feedback control signal u_{fb}^i are to be multiplied with a weighting function $h(\tau)$ obtained through locally symmetrically double-integration (WLSI^2) and multiplication by a learning gain γ.

Remark 9.2.1. As suggested in [155], when ILC starts to have a substantial impact on how control is actually done in industry, it will be the linear based ILC that leads the way. In engineering practice, to design a control system, it is very fundamental to have a linear proximal model $G(s)$ for frequencies below a frequency of interest, say, ω_c. For a feedback controlled system, it is almost sure that its frequency response can be well approximated by that of a linear system, i.e., $G_c(s)$, the closed-loop transfer function. Therefore, at this point, it is understood that, $G(s)$ in Fig. 9.1 is a linear approximation of the *plant* which may be nonlinear.

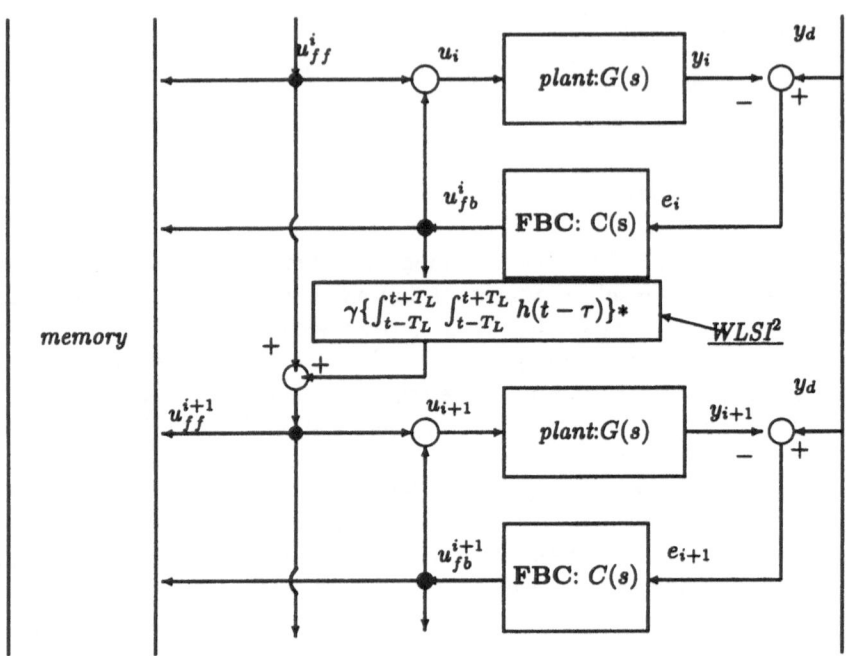

Fig. 9.1. Block Diagram of Weighted Local Symmetrical Double-Integral-type (WLSI^2) Iterative Learning Control

9.2.1 WLSI^2-type ILC Updating Law

The learning updating law is hence written by

$$u_{ff}^{i+1}(t) = u_{ff}^i(t) + \gamma \int_{t-T_{L_1}}^{t+T_{L_1}} h_1(t-\tau) \int_{\tau-T_{L_2}}^{\tau+T_{L_2}} h_2(\tau-\tau_1)u_{fb}^i(\tau_1)d\tau_1 d\tau \quad (9.1)$$

where γ is the learning gain, $h_1(\cdot)$ and $h_2(\cdot)$ are local weighting functions and T_{L_1} and T_{L_2} are the widths of local double-integration. Both $h_1(\cdot)$ and $h_2(\cdot)$ should be chosen according to *locality and symmetry*, i.e.,

$$\begin{cases} h_j(t) = h_j(-t), & \forall t; \\ h_j(t) = 0, & t \notin [-T_{L_j}, T_{L_j}]; \\ h_j(t) \neq 0, & t \in [-T_{L_j}, T_{T_j}], \quad (j = 1, 2). \end{cases} \quad (9.2)$$

Also, $h_2(\cdot)$ should satisfy

$$\int_{-T_{L_j}}^{T_{L_j}} h(\tau)d\tau = 1, \quad (j = 1, 2). \tag{9.3}$$

Clearly, when $T_{L_j} \to 0$, $h_j(\cdot) \to \infty$. The overall control is simply that

$$u_{i+1}(k) = u_{ff}^{i+1}(k) + u_{fb}^{i+1}(k). \tag{9.4}$$

When $T_{L_2} \to 0$, (9.1) becomes

$$u_{ff}^{i+1}(t) = u_{ff}^i(t) + \gamma \int_{t-T_{L_1}}^{t+T_{L_1}} h(t-\tau)u_{fb}^i(\tau)d\tau \tag{9.5}$$

which is simply a weighted local symmetric integral (WLSI) type. In this chapter, we shall consider the case for $T_{L_1} = T_{L_2} = T_L$. Therefore, in Fig. 9.1, there are two parameters - γ the learning gain and T_L the width of the WLSI^2, are to be designed and specified. The corresponding learning algorithm becomes

$$u_{ff}^{i+1}(t) = u_{ff}^i(t) + \gamma \int_{t-T_L}^{t+T_L} h(t-\tau) \int_{\tau-T_L}^{\tau+T_L} h(\tau-\tau_1)u_{fb}^i(\tau_1)d\tau_1 d\tau. \tag{9.6}$$

Here, two examples of the local weighting function $h(t)$ are presented for illustration shown in Fig. 9.2 where the rectangular wave form is denoted by $h_1'(t)$ and the triangular one by $h_2'(t)$. They clearly satisfy conditions (9.2) and (9.3) set for general form of $h(t)$.

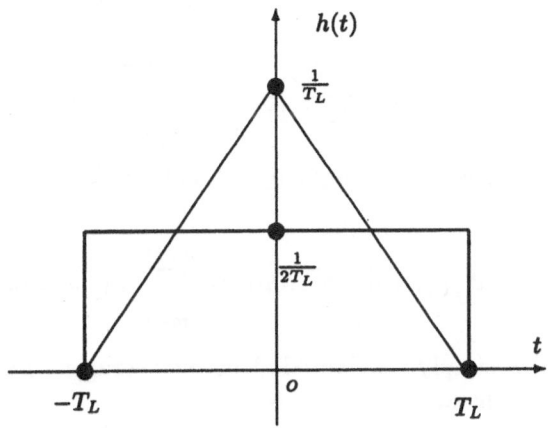

Fig. 9.2. Two simple local weighting functions (rectangular: $h_1'(t)$; triangular: $h_2'(t)$)

Some remarks on the proposed WLSI^2-type ILC scheme are as follows.

Remark 9.2.2. When $T_L \to 0$, learning updating law (9.1) becomes

$$u_{ff}^{i+1}(t) = u_{ff}^i(t) + \gamma u_{fb}^i(t). \tag{9.7}$$

This was discussed in [132].

Remark 9.2.3. **'Anticipatory'** Scheme: When the feedback controller $C(s)$ is of a PD (proportional plus derivative) type, i.e.,

$$C(s) = K_p + K_d s,$$

(9.1) becomes

$$u_{ff}^{i+1}(t) = u_{ff}^i(t) + \gamma K_p' \int_{t-T_L}^{t+T_L} \int_{\tau-T_L}^{\tau+T_L} e^i(\tau_1) d\tau_1 d\tau$$

$$+ \gamma K_d' \int_{t-T_L}^{t+T_L} [e^i(\tau + T_L) - e^i(\tau - T_L)] d\tau,$$

where, $K_p' = K_p/(4T_L^2)$ and $K_d' = K_d/(4T_L^2)$. In the above, we simply assume that $h(t) = h_1'(t)$. Clearly, this is what was noted as *'anticipatory'* scheme [243]. It can be argued that the effectiveness of ILC method is due to the *'anticipatory'* or noncausal [62] usage of the stored data of previous iterations.

Remark 9.2.4. **Discrete-time Scheme**: Consider that the sampling period is T_s. Let $T_L/T_s = M$. Using the simple rectangular formula of the numerical quadrature (Euler approximation), for $h(t) = h_1'(t)$ shown in Fig. 9.2, one obtains

$$u_{ff}^{i+1}(k) = u_{ff}^i(k) + \frac{\gamma}{(2M+1)^2} \sum_{m=-M}^{M} \sum_{j=m-M}^{m+M} u_{fb}^i(k+j). \tag{9.8}$$

For $h(t) = h_2'(t)$, similarly simple updating formula as (9.8) can be obtained. In general,

$$u_{ff}^{i+1}(k) = u_{ff}^i(k) + \gamma \sum_{m=-M}^{M} h(m) \sum_{j=m-M}^{m+M} h(j-m) u_{fb}^i(k+j). \tag{9.9}$$

Clearly, (9.8) is simply a weighted algebraic averager - a special type of non-causal filter discussed in Chapter 8.

Here, our control task is to track the given desired output trajectory $y_d(t)$ over a fixed time interval $[0, T]$ as closely as possible. With an existing feedback controller $C(s)$, the main objective of this chapter is to use a learning feed forward controller (LFFC) given by ILC updating law (9.1) to achieve a better tracking performance. In what follows, we will perform an analysis on ILC convergence and then present a practical procedure for ILC design.

9.3 Convergence Analysis

Before performing the convergence analysis of the proposed learning scheme, we should clarify the existence problem. That is, for a given $y_d(t)$, there exists a unique feedforward $u_{ff}^\infty(t)$ such that $y_\infty(t) \to y_d(t)$ for all $t \in [0, T]$.

The convergence of the proposed learning controller is in the sense that u_{ff}^i approaches to a fixed point signal as i increases and meanwhile, $y_i(t) \to y_d(t)$. This is summarized in the following theorem.

Theorem 9.3.1 *A linear system shown in Fig. 9.1 is controlled by a suitable feedback controller which performs a given task repeatedly. A weighted local symmetrical double-integral-type ILC scheme (9.1) is applied as a learning feedforward controller (LFFC). There exists a real constant γ and a positive $T_L(0 < T_L < T)$ such that the learning process is convergent and furthermore,*

$$\lim_{i \to \infty} U_{ff}^i(s) \to Y_d(s)/G(s). \tag{9.10}$$

where $U_{ff}^i(s) = \mathcal{L}[u_{ff}^i(t)]$ and $Y_d(s) = \mathcal{L}[y_d(t)]$. The convergence rate is given by

$$\rho(\omega, \gamma, T_L) \triangleq |1 - \gamma H(\omega, T_L)G_c(j\omega)| < 1, \tag{9.11}$$

where $G_c(s) = C(s)G(s)/(1 + C(s)G(s))$.

Theorem 9.3.1 implies that the iterative learning controller is essentially applied to invert the plant controlled in an iterative manner. Since linear system is considered in this chapter, in the sequel, frequency domain notion is used. Using Laplace transformation and its convolution theorem, the updating law (9.1) becomes

$$U_{ff}^{i+1}(s) = U_{ff}^i(s) + \gamma H(\omega, T_L)U_{fb}^i(s) \tag{9.12}$$

where $U_{fb}^i(s) = \mathcal{L}[u_{fb}^i(t)]$, $s = j\omega$ and

$$H(\omega, T_L) = \left(\int_{-T_L}^{T_L} h(t)e^{-st}dt\right)^2. \tag{9.13}$$

Due to its symmetry, the phase of $H(\omega, T_L)$ is zero.

Now we proceed to present a proof of Theorem 9.3.1.

Proof. From Fig. 9.1, the feedback signal can be written as

$$U_{fb}(s) = -G_c(s) + G_c(s)Y_d(s)/G(s). \tag{9.14}$$

Learning updating law (9.12) becomes,

$$U_{ff}^{i+1}(s) = [1 - \gamma H(\omega, T_L)G_c(s)]U_{ff}^i(s)$$
$$+ \gamma H(\omega, T_L)G_c(s)Y_d(s)/G(s). \tag{9.15}$$

Iterating (9.15), one obtains

$$U_{ff}^{i+1}(s) = [1 - \gamma H(\omega, T_L)G_c(s)]^i U_{ff}^0(s)$$
$$+ \{1 - [1 - \gamma H(\omega, T_L)G_c(s)]^{i+1}\} Y_d(s)/G(s). \qquad (9.16)$$

Since $H(\omega, T_L)$ and $G_c(s)$ are essentially low pass filters, it is clearly possible to choose a suitable γ such that (9.11) is true. In addition, T_L can be used to shape $\rho(\omega, \gamma, T_L)$ which is the convergence rate in (9.16). Therefore, there exists a design of γ and T_L such that (9.11) holds. From (9.16) $\lim_{i \to \infty} U_{ff}^i(s) \to Y_d(s)/G(s)$ and moreover, $y_i(t) \to y_d(t)$ for all $t \in [0, T]$ as $i \to \infty$.

Remark 9.3.1. From the above proof, it can be seen that the learning convergence is independent of u_{ff}^0 which means that the initial feedforward control can be chosen arbitrarily. However, u_{ff}^0 can be practically set to 0 because no prior knowledge on determining u_{ff}^0 is available. In addition, it is implied in (9.12) that the initial condition of each iteration should be the same.

Remark 9.3.2. For $h_1'(t)$ and $h_2'(t)$ shown in Fig. 9.2, it can be obtained from (9.13) that

$$H_1'(\omega, T_L) = \frac{\sin^2(\omega T_L)}{(\omega T_L)^2}; \qquad (9.17)$$

$$H_2'(\omega, T_L) = \frac{4(1 - \cos(\omega T_L))^2}{\omega^4 T_L^4} = \frac{\sin^4(\omega T_L/2)}{(\omega T_L/2)^4} = H_1'^2(\omega, T_L/2). \quad (9.18)$$

Let $H_1''(\omega, T_L) = \frac{\sin(\omega T_L)}{\omega T_L}$ and $H_2''(\omega, T_L) = \frac{2(1-\cos(\omega T_L))}{\omega^2 T_L^2}$ which are actually frequency responses of zero phase filters. In terms of frequency response, we can see from Fig. 9.3 that $H_1'(\omega, T_L)$ is between $H_1''(\omega, 2T_L)$ and $H_1''(\omega, T_L)$ at the lower frequency band. However, the high frequency response of $H_1'(\omega, T_L)$ are much better than that of $H_1''(\omega, T_L)$. Similar observation can be made for $H_2'(\omega, T_L)$ from Fig. 9.4. Especially, $H_2'(\omega, T_L)$ is very close to $H_1'(\omega, T_L)$ as shown in Fig. 9.4. This is even more suitable for practical applications. Therefore, $H_j'(\omega, T_L)$ is more preferred than $H_j''(\omega, T_L)$ in practical use ($j = 1, 2$). That is, the learning updating law (9.1) is more efficient than (9.5) in handling high frequency uncertainty or disturbance.

9.4 ILC Design Method

As discussed in Chapter 8, the knowledge about the plant to be controlled is important when presenting a control design method. Here, the knowledge we used includes

- The frequency content of the desired trajectory, which is less than a known frequency denoted by ω_d and $\omega_d < \omega_c$ where ω_c is the systems cut-off frequency;

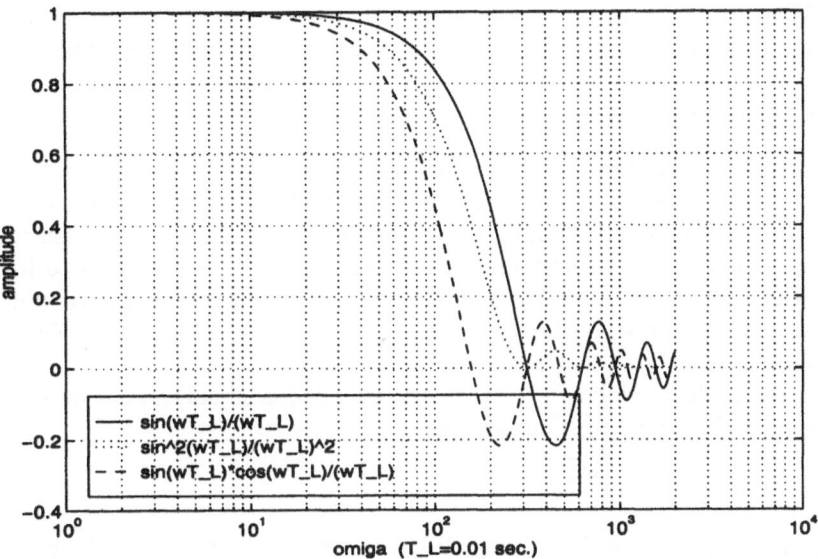

Fig. 9.3. Frequency response comparison of $H_1'(\omega, T_L)$ and $H_1''(\omega, T_L)$ ($T_L=0.01$ sec.)

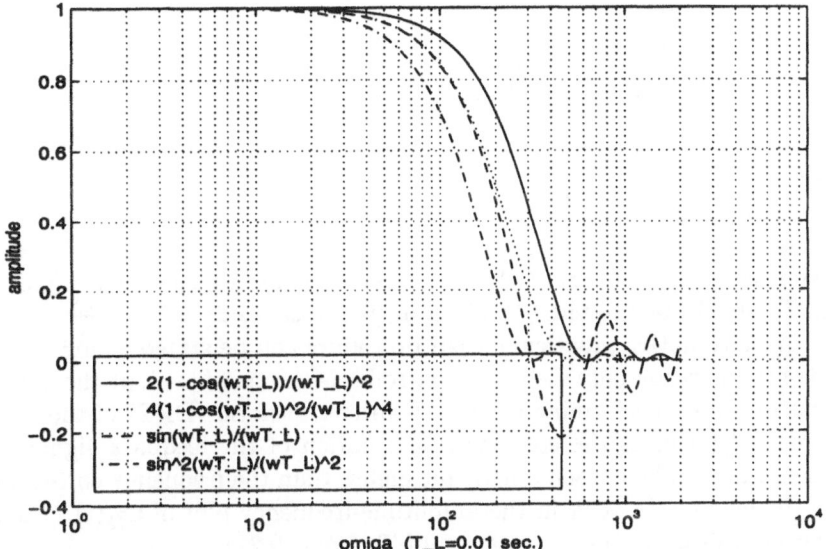

Fig. 9.4. Frequency response comparison of $H_2'(\omega, T_L)$ and $H_2''(\omega, T_L)$ ($T_L=0.01$ sec.)

• An estimate of $G_c(j\omega_c)$ or $G_c(j\omega_d)$.

Clearly, the above knowledge is minimal for controller design. In addition, the local weighting function $h(t)$ for a given T_L should be decided.

9.4.1 Design Method for T_L

The basic intuition in choosing a T_L is that it can neither be too large nor too small. Small T_L will result in more high frequency signal components stored in the memory bank. These high frequency signal components may be accumulated due to different phase relationship from iteration to iteration. Meanwhile, a too large T_L will deteriorate the low frequency components of the signal when smoothing out the high frequency components. Therefore, a suitably chosen T_L is very important.

A simple consideration is that, the energy of a signal can not be attenuated by half via $H(\omega, T_L)$. One obtains

$$H(\omega_d, T_L) = \frac{\sqrt{2}}{2}. \tag{9.19}$$

Using $H_1'(\omega, T_L)$ and applying (9.19) will give an estimate of T_L by

$$\omega_d T_L \approx 1.002 \tag{9.20}$$

using `solve('sin(x)/x =sqrt(sqrt(2)/2)')` of MATLAB Symbolic Math Toolbox. Therefore,

$$T_L \geq \frac{1.002}{\omega_d}. \tag{9.21}$$

Similarly, when using $H_2'(\omega, T_L)$,

$$\omega_d T_L \approx 1.43 \tag{9.22}$$

and

$$T_L \geq \frac{1.43}{\omega_d} \tag{9.23}$$

using `solve('2*(1-cos(x))/x^2 = sqrt(sqrt(2)/2)')`. The upper bound on T_L should be around $\frac{1.43}{\omega_d}$ given in (9.23), say, $1 \pm 10\%$ of $\frac{1.43}{\omega_d}$.

An alternative practical design procedure can be stated as follows. As at the first iteration only feedback controller is commissioned, at the end of the first iteration, performing discrete Fourier transform (DFT) of the feedback signal $u_{fb}^0(t)$ gives the spectrum information of the feedback signal. A frequency ω_c' can be chosen a little bit higher than the frequency corresponding to the magnitude peak in the amplitude-frequency plot of $u_{fb}^0(t)$. Then, one can use this ω_c' to obtain a design of T_L from (9.21).

Remark 9.4.1. The learning scheme analyzed in [132] is actually the case when $T_L \to 0$ which caters for tracking desired trajectory with ultra high signal frequency, according to the discussion of this subsection. This is too stringent to be practically useful. Therefore, in practical use of learning control scheme like the one proposed in [132], a local symmetrical integration is required with a suitable T_L.

9.4.2 Design Method for γ

During practice, it is always possible to start with a smaller, conservative γ via which the learning process converges. Then fine tune of γ is still possible as discussed in Sec. 8.5.4. The above discussions indicate that it is an easy task to make ILC work.

Full knowledge of $G_c(j\omega)$, $\omega \leq \omega_c$ may be sometimes impractical. Therefore, it is assumed that at least the value of $G_c(j\omega_c)$ is available. In what follows, it will be shown that $G_c(j\omega_c)$ can be used to design a reasonable γ. Let $G(j\omega) = A(\omega)e^{\theta(\omega)}$. Denote $\bar{\rho} = \rho^2(\omega, \gamma, T_L)$. Then, from (9.11),

$$\bar{\rho} = 1 - 2\gamma H(\omega, T_L)A(\omega)\cos\theta(\omega) + \gamma^2 H^2(\omega, T_L)A^2(\omega). \qquad (9.24)$$

Clearly, γ should be chosen to minimize $\bar{\rho}$. From (9.24), the best γ should be

$$\gamma = \frac{\cos\theta(\omega)}{H(\omega, T_L)A(\omega)} \qquad (9.25)$$

by setting $\frac{d\bar{\rho}}{d\gamma} = 0$. At frequency ω_c,

$$\gamma = \frac{\cos\theta(\omega_c)}{H(\omega_c, T_L)A(\omega_c)}. \qquad (9.26)$$

It should be noted that γ may be negative at certain frequency range. For most applications, ω_d is quite small and in this case γ can be given approximately by

$$\gamma \leq \sqrt{2}. \qquad (9.27)$$

When only $G_c(j\omega_d)$ is known, γ can be design similarly according to (9.25). If the knowledge of $G_c(j\omega)$ within a frequency range $[\omega_L, \omega_H]$ is available. Then, a plot of $\gamma(\omega)$ can be obtained from (9.26). This plot is useful in selecting a suitable γ when different frequencies of interest are to be considered in $[\omega_L, \omega_H]$.

In any case, it is possible to tune γ to make the ILC convergence as fast as possible. However, as shown in Sec. 9.4.3, the ILC convergence rate has its limit governed by the closed loop system dynamics alone.

9.4.3 ILC Convergence Rate

Substituting (9.25) into (9.24), the corresponding minimal $\bar{\rho}$ is given by

$$(\bar{\rho})_{\gamma,\min} = \sin^2\theta(\omega), \qquad (9.28)$$

i.e.,

$$\rho \geq |\sin\theta(\omega)| = \rho^*.$$

The above inequality implies that, for a given ω of interest, the ILC convergence rate cannot be faster than the limit characterized by ρ^*. This limit is independent of learning schemes applied. The only way to achieve a faster

ILC convergence process is to well design the feedback controller $C(j\omega)$ such that the phase response of the closed-loop system will well behave as required in (9.28).

Similarly, certain heuristic designs as discussed in Chapter 8 can also be applied here.

9.5 Conclusions

A new iterative learning control (ILC) updating law is proposed for tracking control of continuous linear system over a finite time interval. The ILC is applied as a feedforward controller to the existing feedback controller. By a weighted local symmetrical double-integration of the feedback control signal of the previous iteration, the ILC updating law takes a simple form with only two design parameters: the learning gain and the range of local double-integration. Convergence analysis is presented together with a design procedure. Also, the ILC convergence rate has been shown to be limited by the characteristics of the feedback controlled system.

10. Iterative Learning Identification with an Application to Aerodynamic Drag Coefficient Curve Extraction Problem

MOST OF THE ILC APPLICATIONS are found in robot manipulator control. In this chapter, by applying the results of previous chapters, we present a new application of ILC for identifying (extracting) the projectile's optimal fitting drag coefficient curve C_{df} from radar measured velocity data. This is considered as an optimal tracking control problem (OTCP) where C_{df} is regarded as a *virtual control* while the radar measured velocity data are taken as the *desired output trajectory* to be optimally tracked. With a 3-DOF point mass trajectory prediction model, a high-order iterative learning identification scheme with time varying learning parameters is proposed to solve this OTCP. It should be pointed out that the performance index of this OTCP is in a *minimax* sense and the method works for an arbitrarily chosen initial curve (control function) – C_{df}.

To compare the effectiveness of the proposed ILC method applied to an aerodynamic coefficient curve identification problem of this Chapter, an alternative but less effective scheme – Optimal Dynamic Fitting method is compared. The results obtained by these two methods are shown to be *cross-validatedable*. The advantages of ILC method for such an application are summarized in this Chapter.

10.1 Introduction

In exterior ballistics, projectile's drag coefficient curve C_{df} plays a key role in applications such as the firing table generation and so on. It can be obtained from wind tunnel testing, engineering prediction, and theoretical calculation. The accuracy of the obtained C_{df} may suffer from various factors such as the interferences between projectile (maybe its model) and wind tunnel walls, the assumptions imposed in its numerical calculation which are sometimes not identical to realistic conditions, and etc. Therefore, identifying the aerodynamic properties of real or full-scale flying objects from flight testing data has been an important research area in modern flight dynamics [38, 150, 23, 246, 93, 85, 14, 121, 47, 152]. This is in essence a combination of modern control theory and flight dynamics. It is a common belief that the curve extracted directly from actual flight testing data is more practical and can be applied in the verification and improvement of design objectives, in validation of computational aerodynamic property prediction codes, and so on. The problem considered in this chapter is to extract projectile's drag coefficient curve from Doppler tracking radar measured velocity data.

In most of the proving grounds, the Doppler tracking radar such as TERMA OPOS DR-582 has become a popular equipment. The velocity data of flight tests measured by the tracking radar will be used in the identification of a single aerodynamic drag coefficient curve of an artillery projectile. It should be pointed out that, from the aerodynamic point of view, the identified $C_{df}(M)$ curve cannot be taken as a zero lift drag coefficient curve or an incidence induced drag coefficient curve. The $C_{df}(M)$ is just the *fitting drag coefficient curve* with respect to the trajectory model employed. Hence, $C_{df}(M)$ comprehensively reflects the effects of zero lift drag and the drag induced by the angular motion around the center of mass. As the generation of firing table is mainly based on a single ballistic coefficient, a drag law and some fitting factors, thus more accurate firing table can be produced when the $C_{df}(M)$ identified is utilized.

In many practical system identification tasks, identification of parameters is actually a special case of nonlinear function or curve identification. On the other hand, curve identification or curve extraction from system I/O data can be easily converted to parameter identification through a parameterization procedure of the curve to be identified [47, 167, 52]. Identifying a curve in causal systems can often be regarded as an optimal tracking control problem (OTCP) where the observed system outputs are the desired trajectories to be optimally tracked. The OTCP might be a singular optimal control problem (SOCP) in some applications and thus difficulties will arise in numerical solution of SOCP [47, 52]. In this chapter, a new method – "iterative learning" is proposed for identifying a nonlinear function in a nonlinear dynamic system, where the testing data are considered as the desired trajectory. Through learning iterations, the infinite norm of the tracking error tends to be minimum asymptotically.

By using the iterative learning identification, we can have the following major advantages. Firstly, the method is very simple and straightforward for an easy implementation with less computation efforts. Secondly, the asymptotic convergence condition is easy to be satisfied and appropriate learning parameters are easy to be determined. Thirdly, the convergence is in a global sense, i.e., the convergence is not sensitive to initial curve trial. Finally, it is worth noticing that the performance index is in a form equivalent to minimax one, which is more practical in engineering practice.

The remaining parts of this chapter are organized as follows. The curve identification problem is formulated in Sec. 10.2. The general idea and procedures for the iterative learning curve identification are presented in Sec. 10.3. A convergence analysis of the proposed high-order iterative learning scheme is given in a general form in Sec. 10.4. Practical issues in the determination of learning parameters are presented in Sec. 10.5. Additionally, a bi-linear

ILC scheme is proposed in Sec. 10.6 and the improved results are given in order to stimulate the nonlinear ILC research. In Sec. 10.7, based on a set of actual flight testing data, learning identification results are compared for several schemes given in Sec. 10.3. Sec. 10.8 concludes this chapter.

10.2 A Curve Identification Problem

For brevity of our discussion, a 3-DOF point mass ballistic model is used. A more complete ballistic model [43] may be used if other aerodynamic coefficients are available and the method presented here can still be applied. Suppose at time t, the position of the projectile P' in earth coordinate system (ECS) is $[x(t), y(t), z(t)]^T$ and its relative velocity vector $\vec{u}$ w.r.t. ECS is $[u_x(t), u_y(t), u_z(t)]^T$. The position of the radar R in ECS is $[x_r(t), y_r(t), z_r(t)]^T$, which is known as illustrated in Fig. 10.1. The 3-DOF point mass trajectory model can be described by nonlinear state space equations as follows.

$$\begin{cases} \dot{u}_x(t) = -\rho_a(t)sV(t)(u_x(t) - w_x)C_{df}(t)/2m_0 \\ \dot{u}_y(t) = -\rho_a(t)sV(t)u_y(t)C_{df}(t)/2m_0 - g_0(t) \\ \dot{u}_z(t) = -\rho_a(t)sV(t)(u_z(t) - w_z(t))C_{df}(t)/2m_0 \\ \dot{x}(t) = u_x(t), \quad \dot{y}(t) = u_y(t), \quad \dot{z}(t) = u_z(t) \end{cases} \tag{10.1}$$

where $t \in [t_0, T]$ and t_0, T are known, g_0 is the gravitational acceleration, w_x, w_z are the wind components in ECS known from meteorological measurements, V is projectile's relative velocity w.r.t. the wind and

$$V = \sqrt{(u_x - w_x)^2 + u_y^2 + (u_z - w_z)^2}, \tag{10.2}$$

ρ_a is the air density, $s = \pi d^2/4$ is the reference area of the projectile and d is the projectile's reference diameter, m_0 is the mass of the projectile, and $C_{df}(t)$ is the fitting drag coefficient curve w.r.t. the trajectory model (10.1), which is regarded as the *virtual* unconstrained control function.

Define a state vector $X(t)$ of system (10.1) as $[u_x(t), u_y(t), u_z(t), x(t), y(t), z(t)]^T$ and the initial state $X(t_0)$ is known from the flight testing setup. Also let M denote the Mach number, i.e. $M \triangleq V/a$, where a is the local sonic speed. $C_{df}(M)$ can be composed from $C_{df}(t)$ and $M(t)$, which is more concerned from the aerodynamics point of view.

Referring to Fig. 10.1, let $\bar{r}(t) = \overline{R'P'}$ represent the distance between the tracking radar R' and the projectile P'. Then,

$$\dot{v}_d(t) = \dot{r}(t). \tag{10.3}$$

Denote $v_r(t)$ the Doppler radar measured velocity data. To formulate our problem, we must transform the projectile's tangential velocity $\vec{u}$ into a velocity v_d in the radial direction of the Doppler radar as follows. Let

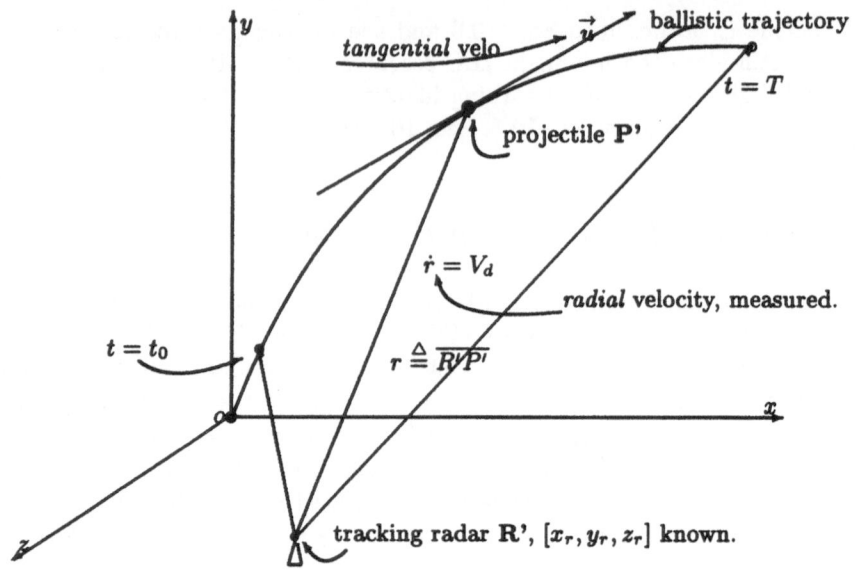

Fig. 10.1. Illustration of Doppler Radar Tracking

$$[r_x, r_y, r_z] = [x - x_r, y - y_r, z - z_r], \tag{10.4}$$

$$\bar{r} = \sqrt{r_x^2 + r_y^2 + r_z^2}.$$

Then, the 'output equation' is that

$$v_d = (u_x r_x + u_y r_y + u_z r_z)/\bar{r}. \tag{10.5}$$

Denote $e(t) = v_r(t) - v_d(t)$ the tracking error. Then, the control problem is to achieve the following objective:

$$\min_{C_{df}(t)} e_b \triangleq \min_{C_{df}(t)} \sup_{t \in [0,T]} |e(t)|. \tag{10.6}$$

It can be noted that (10.1), (10.5), and (10.6) formulate an OTCP. In this OTCP, the Doppler radar measured velocity data $v_r(t)$ is taken as the desired trajectory to be tracked by 'output' v_d and the fitting drag coefficient curve $C_{df}(t)$ is the 'control' function.

However, a lot of difficulties will arise in numerical computation for this OTCP with a minimax performance index (10.6). Even when linear quadratical (LQ) performance index is used, the OTCP is obviously singular as pointed out in [52] and thus the mathematical programming based techniques [44] are impractical both in computing cost and in convergence property. We shall show that, in the next section, this OTCP can be easily solved by iterative learning concept.

10.3 Iterative Learning Identification

10.3.1 Basic Ideas

We have discussed iterative learning control in previous chapters. The concept of iterative learning control can also be applied for curve identification. A illustrative diagram is shown in Fig. 10.2. In Fig. 10.2, $y_d(t)$ is the given measurement data (desired trajectory); $e_i(t)$ is the estimation error (tracking error) at the i-th repetition; $u_i(t)$ is the curve to be identified. After the i-th repetition, $u_i(t)$ and $e_i(t)$ are to be stored in the memory pool for the use in the next iteration.

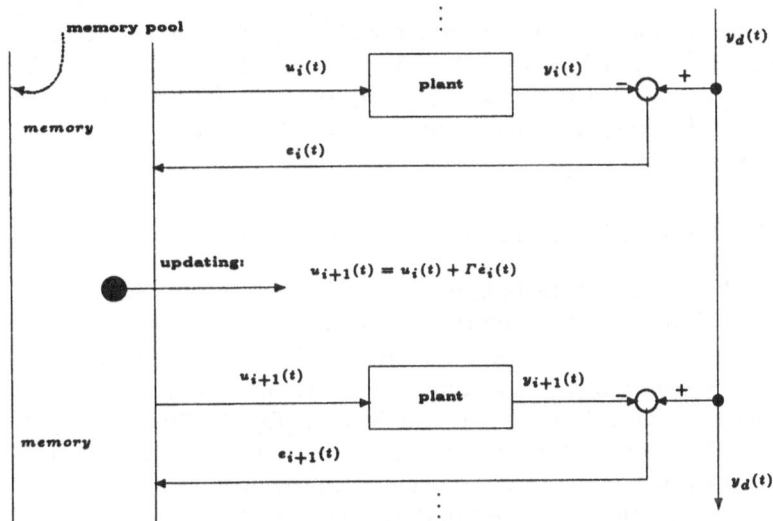

Fig. 10.2. Block-Diagram of Iterative Learning *Control* or *Curve Identification*

The differences between iterative learning control and iterative learning identification are as follows:

- *iterative learning control*: given a desired output trajectory, to solve a desired function iteratively along with the system repetitive operations;
- *iterative learning identification*: to identify an unknown nonlinear function, which is regarded as a virtual control function, from I/O data, which is taken as the desired output trajectory, iteratively.

10.3.2 Learning Identification Procedures

The key issue in implementing an iterative learning identification method is how the *system* is operated repeatedly. In fact, one system repetition here is simply to numerically integrate the trajectory model (10.1) from t_0 to T with initial condition $X(t_0)$ under the current control $C_{df}(t)$. Then, based on Fig. 10.2, the iterative identification procedures can be summarized as follows:

- *Step 1.* Set $i = 0$ and give an arbitrary $(C_{df}(t))_0$. Specify a maximum error ε^* allowed and a maximum number of iteration N_i for identifications.
- *Step 2.* Integrate the ballistic model (10.1) with $(C_{df}(t))_i$ from t_0 to T with initial condition $X(t_0)$. Then obtain the tracking error $e_i(t)$ according to (10.5).
- *Step 3.* Learning updating (referring to Fig. 10.2)

$$(C_{df}(t))_{i+1} = (C_{df}(t))_i + r_v \dot{e}_i(t) \tag{10.7}$$

  where r_v is a learning gain which will be specified in Sec. 10.5.
- *Step 4.* If either $e_b < \epsilon^*$ or $k \geq N_i$ then goto *Step 6*; else goto *Step 5*.
- *Step 5.* $i = i + 1$ and store $(C_{df}(t))_i$ and $\dot{e}_i(t)$ for possible use such as in *high-order* learning schemes given in Sec. 10.3.3. Goto *Step 2*.
- *Step 6.* Stop.

10.3.3 High-order Iterative Learning Scheme

From the first-order learning updating law as shown in Fig. 10.2, it is quite intuitive that the more information of more past iterations are used, the more learning performance improvements can be expected. This is called the *high-order learning updating law*, which was first proposed in [27] and further investigated in [49]. As explained in [46], the first-order updating law is actually a pure integral controller in the iteration number axis. Hence one can expect that a high-order scheme can provide better performance.

In general, the high-order updating law can be written as follows:

$$(C_{df}(t))_{i+1} = \sum_{k=1}^{N} [p_k(t)(C_{df}(t))_l + r_k(t)\dot{e}_l(t)] \tag{10.8}$$

where i is the ILC iteration number, i.e., the cycle number; p_k, r_k are the time varying learning gains; N the order of ILC updating law and $l \overset{\triangle}{=} i - k + 1$. The *Learning Identification* procedures for implementation are similar to those given in Sec. 10.3.2. An ILC convergence analysis is given in detail in Sec. 10.4 in a general setting, where some conditions on the learning parameters p_k, r_k for convergence are established in Theorem 10.4.1. The determination of p_k, r_k is given in Sec. 10.5.

10.4 Convergence Analysis

For the learning convergence analysis in a general sense, consider a class of repetitive uncertain nonlinear time-varying system

$$\begin{cases} \dot{\bar{x}}_i(t) = f(\bar{x}_i(t), t) + B(\bar{x}_i(t), t)\bar{u}_i(t) + \bar{w}_i(t) \\ \bar{y}_i(t) = g(\bar{x}_i(t), t) + \bar{v}_i(t) \end{cases} \tag{10.9}$$

where $t \in [0, T]$ and T is given; i denotes the i-th repetition of the system operation; $\bar{x}_i(t) \in R^n$, $\bar{u}_i(t) \in R^m$, and $\bar{y}_i(t) \in R^r$ are the state, control input, and output of the system, respectively; the functions $f(\cdot, \cdot)$, $B(\cdot, \cdot) : R^n \times [0, T] \mapsto R^n$, and $g(\cdot, \cdot) : R^n \times [0, T] \mapsto R^r$ are piecewise continuous and satisfy the Lipschitz continuity conditions, i.e., $\forall t \in [0, T]$,

$$\|f(\bar{x}_{i+1}(t), t) - f(\bar{x}_i(t), t)\| \le k_f(\|\bar{x}_{i+1}(t) - \bar{x}_i(t)\|),$$

$$\|B(\bar{x}_{i+1}(t), t) - B(\bar{x}_i(t), t)\| \le k_B(\|\bar{x}_{i+1}(t) - \bar{x}_i(t)\|),$$

$$\|g(\bar{x}_{i+1}(t), t) - g(\bar{x}_i(t), t)\| \le k_g(\|\bar{x}_{i+1}(t) - \bar{x}_i(t)\|),$$

where k_f, k_B, $k_g > 0$ are the Lipschitz constants; $\bar{w}_i(t)$, $\bar{v}_i(t)$ are uncertainty or disturbance to the system with bounds $b_{\bar{w}}, b_{\bar{v}}$ defined as

$$b_{\bar{w}} \triangleq \sup_{t \in [0, T]} \|\bar{w}_i(t)\|, \quad b_{\bar{v}} \triangleq \sup_{t \in [0, T]} \|\bar{v}_i(t)\|, \, \forall i,$$

$$g_{\bar{x}i} \triangleq \partial g(\bar{x}_i(t), \cdot)/\partial \bar{x}_i(t), \quad g_{ti} \triangleq \partial g(\bar{x}_i(t), \cdot)/\partial t.$$

The proposed high-order D-type ILC updating law is

$$\bar{u}_{i+1}(t) = \sum_{k=1}^{N} [P_k(t)\bar{u}_l(t) + R_k(t)\dot{e}_l(t)] \tag{10.10}$$

where $l \triangleq i - k + 1$, N the order of ILC updating law; $P_k(t), R_k(t)$ are bounded time-varying learning parameter matrices of proper dimensions; $e_i(t) \triangleq \bar{y}_d(t) - \bar{y}_i(t)$ is the tracking error and $\bar{y}_d(t)$ is the realizable desired system output. The realizable $\bar{y}_d(t)$ means that there exists a unique bounded $\bar{u}_d(t), \bar{x}_d(t)$ such that

$$\begin{cases} \dot{\bar{x}}_d(t) = f(\bar{x}_d(t), t) + B(\bar{x}_d(t), t)\bar{u}_d(t) \\ \bar{y}_d(t) = g(\bar{x}_d(t), t) \triangleq g_d. \end{cases} \tag{10.11}$$

Denote $\dot{\bar{x}}_d(t) \triangleq f_d + B_d \bar{u}_d$. As the initial state at each ILC operation may not be the same, i.e., $\bar{x}_i(0) \ne \bar{x}_d(0)$, we assume the bounded initialization errors as $\|\bar{x}_d(0) - \bar{x}_i(0)\| \le b_{\bar{x}_0}$ and for brevity, we denote $b_{\bar{u}_d} \triangleq \sup_{t \in [0, T]} \|\bar{u}_d\|$, $b_{\dot{\bar{x}}_d} \triangleq \sup_{t \in [0, T]} \|\dot{\bar{x}}_d\|$.

We will show that under certain conditions on the time-varying learning matrices $P_k(t), R_k(t), k = 1, 2, \cdots, N$, the ILC will converge in such a way that when starting from an arbitrary continuous initial control input $\bar{u}_0(t)$, through the ILC updating law (10.10), $\bar{y}_i(t) \to \bar{y}_d(t) \pm \varepsilon^*$ as $i \to \infty$ where the tracking error bound ε^* is related to the uncertainty, disturbance and initialization error. A sufficient condition for ILC convergence is given in the following theorem.

Theorem 10.4.1 *Consider the learning updating law (10.10) applied to the repetitive nonlinear uncertain system (10.9) with bounded uncertainty, disturbance and initialization error. For a given realizable desired trajectory $\bar{y}_d(t)$ on the fixed time interval $[0,T]$, under the condition that*

(C1): $\sum_{k=1}^{N} P_k(t) = I_m$,
(C2): *there exist $\bar{\rho}_k$ satisfying*

$$\|P_k(t) - R_k(t)g_x B(\bar{x}(t),t)\| \le \bar{\rho}_k,$$

for all $(\bar{x}(t),t) \in R^n \times [0,T]$, and

$$\sum_{k=1}^{N} \bar{\rho}_k \stackrel{\triangle}{=} \bar{\rho} < 1, \tag{10.12}$$

then, when $i \to \infty$, the bounds of the tracking errors $\|\bar{u}_d(t) - \bar{u}_i(t)\|$, $\|\bar{x}_d(t) - \bar{x}_i(t)\|$ and $\|\bar{y}_d(t) - \bar{y}_i(t)\|$ converge asymptotically to a residual ball centered at the origin. Additionally, when the bounds of the initialization error, disturbance and uncertainty tend to zero, all the bounds of the tracking errors tend to zeroes, too.

Proof. For brevity, we will drop the time t in the equations. From (10.9), it is easy to see that

$$e_l \stackrel{\triangle}{=} \bar{y}_d - \bar{y}_l = \delta g_l - \bar{v}_l \tag{10.13}$$

where $\delta g_l \stackrel{\triangle}{=} g_d - g(\bar{x}_l(t),t)$;

$$\begin{aligned}
\dot{e}_l = \dot{\bar{y}}_d - \dot{\bar{y}}_l &= g_{xd}\dot{\bar{x}}_d + g_{td} - g_{xl}\dot{\bar{x}}_l - g_{tl} - \dot{\bar{v}}_l \\
&= \delta g_{xl}\dot{\bar{x}}_d + g_{xl}\delta\dot{\bar{x}}_l + \delta g_{tl} - \dot{\bar{v}}_l
\end{aligned} \tag{10.14}$$

where $g_{xd} \stackrel{\triangle}{=} \frac{\partial g(\bar{x}_d(t),t)}{\partial \bar{x}_d(t)}$, $g_{td} \stackrel{\triangle}{=} \frac{\partial g(\bar{x}_d(t),t)}{\partial t}$, $\delta g_{xl} \stackrel{\triangle}{=} g_{xd} - g_{xl}$, $\delta\dot{\bar{x}}_l \stackrel{\triangle}{=} \dot{\bar{x}}_d - \dot{\bar{x}}_l$, and $\delta g_{tl} = g_{td} - g_{tl}$. Similarly, we have

$$\delta\dot{\bar{x}}_l = \delta f_l + \delta B_l \bar{u}_d + B_l \delta\bar{u}_l - \bar{w}_l, \tag{10.15}$$

where $f_l \stackrel{\triangle}{=} f(\bar{x}_l(t),t), B_l \stackrel{\triangle}{=} B(\bar{x}_l(t),t), \delta B_l \stackrel{\triangle}{=} B_d - B_l, \delta f_l \stackrel{\triangle}{=} f_d - f_l, \delta\bar{u}_l \stackrel{\triangle}{=} \bar{u}_d - \bar{u}_l$.

Considering $\delta\bar{u}_{i+1}$ and referring to **(C1)**, (10.10), (10.14) and (10.15), we observe that

$$\begin{aligned}
\delta\bar{u}_{i+1} &= \bar{u}_d - \sum_{k=1}^{N}[P_k(t)\bar{u}_l(t) + R_k(t)\dot{e}_l(t)] \\
&= \sum_{k=1}^{N}(P_k\delta\bar{u}_l - R_k\dot{e}_l) = \sum_{k=1}^{N} P_k\delta\bar{u}_l \\
&\quad - \sum_{k=1}^{N} R_k(\delta g_{xl}\dot{\bar{x}}_d + g_{xl}\delta\dot{\bar{x}}_l + \delta g_{tl} - \dot{\bar{v}}_l)
\end{aligned}$$

$$= \sum_{k=1}^{N}(P_k - R_k g_{\bar{x}l} B_l)\delta \bar{u}_l$$

$$- \sum_{k=1}^{N} R_k[\delta g_{\bar{x}l}\dot{\bar{x}}_d + \delta g_{tl} - \dot{\bar{v}}_l$$

$$+ g_{\bar{x}l}(\delta f_l + \delta B_l \bar{u}_d - \bar{w}_l)] \tag{10.16}$$

Estimating the norms of (10.16), we obtain

$$\|\delta \bar{u}_{i+1}\| \le \sum_{k=1}^{N} \bar{\rho}_k \|\delta \bar{u}_l\| + \sum_{k=1}^{N} \bar{\eta}_k \|\delta \bar{x}_l\| + \varepsilon_1 \tag{10.17}$$

where $\bar{\eta}_k \triangleq b_{R_k}[k_{g_{\bar{x}}}b_{\dot{\bar{x}}d} + k_{g_t} + b_{g_{\bar{x}}}(k_f + k_B b_{\bar{u}_d})]$, $b_{R_k} \triangleq \sup_{t\in[0,T]}\|R_k(t)\|$, and $\varepsilon_1 \triangleq \sum_{k=1}^{N} b_{R_k}(b_{\dot{v}} + b_{\bar{w}}b_{g_{\bar{x}}})$. Integrating (10.15) and then taking the norm yield

$$\|\delta \bar{x}_l\| \le b_{\bar{x}_0} + b_{\bar{w}}T + \int_0^t (k_f + k_B b_{\bar{u}_d})\|\delta \bar{x}_l(\tau)\|\mathrm{d}\tau$$

$$+ \int_0^t b_B \|\delta \bar{u}_l(\tau)\|\mathrm{d}\tau. \tag{10.18}$$

Noticing the fact that

$$\int_0^t \|\delta \bar{x}_l(\tau)\|e^{-\lambda t}\mathrm{d}\tau \le \|\delta \bar{x}_l(t)\|_\lambda \int_0^t e^{-\lambda(t-\tau)}\mathrm{d}\tau$$

$$\le \|\delta \bar{x}_l(t)\|_\lambda O(\lambda^{-1}), \tag{10.19}$$

where $O(\lambda^{-1}) \triangleq (1 - e^{-\lambda T})/\lambda \le 1/\lambda$, performing the λ-norm operation for (10.18), we can obtain

$$\|\delta \bar{x}_l\|_\lambda \le O_1(\lambda^{-1})\|\delta \bar{u}_l\|_\lambda + \varepsilon_1' \tag{10.20}$$

where $O_1(\lambda^{-1}) \triangleq b_B O(\lambda^{-1})/[1 - O(\lambda^{-1})(k_f + k_B b_{\bar{u}_d})]$ and $\varepsilon_1' \triangleq (b_{\bar{x}_0} + b_{\bar{w}}T)/[1 - O(\lambda^{-1})(k_f + k_B b_{\bar{u}_d})]$, if we choose a sufficiently large λ such that

$$O(\lambda^{-1})(k_f + k_B b_{\bar{u}_d}) < 1. \tag{10.21}$$

Taking λ-norm of (10.17) with the substitution of (10.20), we simply have

$$\|\delta \bar{u}_{i+1}\|_\lambda \le \sum_{k=1}^{N} \bar{\rho}'_k \|\delta \bar{u}_l\|_\lambda + \varepsilon_0 \tag{10.22}$$

where $\bar{\rho}'_k \triangleq \bar{\rho}_k + \bar{\eta}_k O_1(\lambda^{-1})$ and $\varepsilon_0 \triangleq \sum_{k=1}^{N} \bar{\eta}_k \varepsilon_1' + \varepsilon_1$. Based on (C2), we can make

$$\bar{\rho}' \triangleq \sum_{k=1}^{N} \bar{\rho}'_k < 1 \tag{10.23}$$

by using a sufficiently large λ. Apply Lemma 2.3.1, we have

$$\lim_{i \to \infty} \|\delta \bar{u}_i\|_\lambda \leq \varepsilon_0/(1 - \bar{\rho}'). \tag{10.24}$$

From (10.20) and (10.13) we get

$$\lim_{i \to \infty} \|\delta \bar{x}_i\|_\lambda \leq \varepsilon_1' + O_1(\lambda^{-1})\varepsilon_0/(1 - \bar{\rho}'), \tag{10.25}$$

$$\lim_{i \to \infty} \|e_i\|_\lambda \leq b_{\bar{v}} + k_g[\varepsilon_1' + \frac{O_1(\lambda^{-1})\varepsilon_0}{(1 - \bar{\rho}')}]. \tag{10.26}$$

Clearly, if the bounds of the uncertainty, disturbance, and initialization error tend to zero, the final tracking error bound will also tend to zero. This completes the proof of Theorem 10.4.1.

10.5 Learning Parameters Determination

The proper determination of learning parameters is critical to the learning convergence performance. Theorem 10.4.1 puts some conditions on the learning gains. This will facilitate the learning parameters determination.

Consider the first order case (10.7) where $N = 1$ and $p_1 = 1$. The best choice of r_v is to make $\bar{\rho} = 0$ which gives fastest learning convergence. To obtain the best choice of r_v, by comparing (10.1) and (10.9), one obtains

$$f(\cdot, \cdot) = [0, -g_0, 0, u_x, u_y, u_z]^T,$$

$$B(\cdot, \cdot) = -\frac{\rho_a s V}{2m}[u_x - w_x, u_y, u_z - w_z, 0, 0, 0]^T,$$

$$g_x(\cdot, \cdot) = \frac{1}{\bar{r}}[r_x, r_y, r_z, u_x - \frac{r_x^2}{\bar{r}^2}u_x, u_y - \frac{r_x^2}{\bar{r}^2}u_y,$$

$$u_z - \frac{r_x^2}{\bar{r}^2}u_z].$$

Therefore, according to (10.12),

$$\bar{\rho} = | 1 + r_v \frac{\rho_a s}{2m} V[v_d - \frac{r_x w_x}{\bar{r}} - \frac{r_z w_z}{\bar{r}}] |, \tag{10.27}$$

Set $\bar{\rho} = 0$, one can get

$$r_v(t) = -\{\frac{\rho_a s}{2m}V[v_d - \frac{r_x w_x}{\bar{r}} - \frac{r_z w_z}{\bar{r}}]\}^{-1}. \tag{10.28}$$

Because $w_x \ll v_d, w_z \ll v_d$, (10.28) can be simplified as

$$r_v(t) \doteq -(\frac{\rho_a s}{2m}V v_d)^{-1}. \tag{10.29}$$

Notice that in the above ideal case, ρ_a, V and v_d will vary with respect to iteration number i and cannot be directly applied. However, (10.29) gives a guideline to select r_v. Based on this and the application of Theorem 10.4.1 with practical considerations, the following high-order ILC updating law with time-varying learning parameters is employed in the learning identification.

$$(C_{df}(t))_{i+1} = \sum_{k=1}^{N} [p_k(C_{df}(t))_l + r_k r^*(t)\dot{e}_l(t)] \tag{10.30}$$

where $l = i - k + 1$; p_k, r_k are constants;

$$r^*(t) \triangleq -2m/(\rho_a^{(0)} s v_r^2) \tag{10.31}$$

and $\rho_a^{(0)}$ is the standard air density at the sea level which can be easily determined. If $r^*(t) = 1$ and $N = 1$, this will be the constant learning parameter case with the first-order updating law (10.7) discussed in [45, 74].

If the learning curve $r^*(t)$ is chosen to vary with respect to time and iteration number simultaneously, better learning performance can be expected. One method is to set

$$(r_v(t))_{i+1} = -[\frac{(\rho_a)_i s}{2m}(V)_i v_r]^{-1}. \tag{10.32}$$

A more simplified version is to set

$$(r_v(t))_{i+1} = -[\frac{(\rho_a)_i s}{2m} v_r^2]^{-1}. \tag{10.33}$$

10.6 A Bi-linear Scheme

As we know, in general, the ILC updating laws such as (10.7) and (10.8) should be in a nonlinear form

$$(C_{df}(t))_{i+1} = \mathbf{L}((C_{df}(t))_i, \ \dot{e}_i(t)) \tag{10.34}$$

where $\mathbf{L}$ is a nonlinear learning mapping. The one shown in Fig. 10.2 is a special linear case of $\mathbf{L}$. As explained in [48], from the idea of Taylor series expansion, a bilinear term should be included in the updating law (10.7) for a possible better learning performance. The nonlinear learning updating law in the ILC research did not draw any attention up to now. This chapter tries to give some practical results to motivate the future theoretical analysis.

The bi-linear learning updating law proposed is as follows:

$$\begin{aligned} (C_{df}(t))_{i+1} &= (C_{df}(t))_i + r_v \dot{e}_i(t) \\ &\quad + \beta(C_{df}(t))_i \dot{e}_i(t) \end{aligned} \tag{10.35}$$

where β is the bi-linear learning gain chosen by trials. As shown in the next section, the bi-linear scheme has some attractive and interesting behaviors.

10.7 Curve Identification Results From Flight Tests

The results presented here are based on a set of real flight testing data. To investigate and compare the curve identification results under different situations, various choices of learning parameters (p_1, p_2, r_1, r_2), as listed in the right-upper corner of each figure, are considered. Clearly, when $p_2 = 0$ and $r_2 = 0$, this is the case with first-order learning updating law, i.e. $N = 1$. If $p_2 \neq 0$ or $r_2 \neq 0$, a high-order learning scheme is employed.

The complete flight testing data is given in the following:

- 1). Projectile's Physical Parameters: d=0.155 m., m=44.99 kg;
- 2). Atmosphere: ICAO Standard, $w_x = w_z = 0.0$ m/sec.;
- 3). Initial State: $(X_0)_i = [480.03, 447.38, 0, 274.6, 410.7, 0]^T$;
- 4). Radar Position: $[x_r, y_r, z_r] = [-67.4, 150.0, 0]$ m.;
- 5). Radar Measured Data: (Note: the v_r series (in equal time interval of $h = 0.3$ second over time period $[0.6, 30.6]$ second) are given in Table 10.1.)

Table 10.1. Radar measured velocity series (m./s., equispaced)

652.9757	644.2016	635.4464	627.0872	618.8109	610.3392
602.2492	594.5349	586.8960	579.4882	572.0804	565.0996
557.9661	551.2083	544.5161	538.0422	531.5992	525.3740
519.2748	513.2719	507.3897	501.6705	495.9301	490.4462
484.9349	479.6691	474.3199	469.1093	464.0607	459.0376
454.1041	449.2432	444.4515	439.6902	435.0269	430.4870
426.0405	421.5356	417.1813	412.8641	408.7190	404.4716
400.4320	396.3070	392.3452	388.3574	384.5102	380.7192
376.9328	373.1922	369.5753	365.9011	362.3417	358.8414
355.2994	351.8651	348.4608	345.0592	341.7309	338.4507
335.2063	331.9882	328.8130	319.5674	316.5649	313.5905
310.6943	307.7371	304.9160	302.0498	299.2961	296.4955
293.7404	291.0726	288.3348	285.7450	283.1665	280.6173
278.1360	275.9500	273.6631	271.5456	269.4207	267.2413
265.2183	263.2231	261.2731	259.3871	257.4787	255.6622
253.8853	252.1319	250.3103	248.6151	246.9184	245.2147
243.5340	241.9693	240.3281			

The identification is studied and the results are presented in five different cases. In each case, the RK-4 method is used to numerically integrate the trajectory model with step size $h = 0.3$ second and a maximal allowable absolute tracking error e_b is set to be 0.05 meter per second for the termination of the identification program. Except in **Case 3**, in all cases, $(C_{df}(t))_0 = 0$.

Case 1: Identification Results with Constant Learning Parameters

In this situation, $r^*(t) = 1$. Fig. 10.3 shows the convergence of tracking error bound e_b in the iterative direction (locally zoomed) for both $N = 1$

and $N = 2$. From the results, it can be noted that the second-order term r_2 improves the convergence significantly.

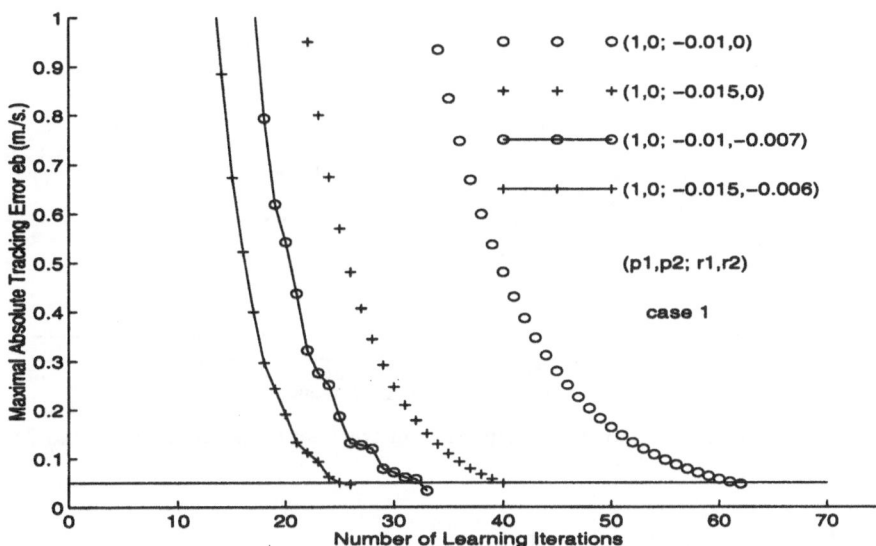

Fig. 10.3. Convergence history comparison of tracking error bound e_b (Case 1)

Case 2: Identification Results with Time-varying Learning Parameters

The time-varying scheme ((10.30), (10.31)) is used with $N = 2$ and with different choices of p_1, p_2, r_1, r_2. It is clear from Fig. 10.4 that the second-order updating law with time varying learning parameters gives a much better convergence. It can also be noted that the convergence of the first order with time varying learning is better than that of both $N = 1$ and $N = 2$ in Fig. 10.3.

Case 3: Identification Results with Different Selections of Initial Curves

To demonstrate that the iterative learning identification method converges in a global sense, results from different choices of initial curve $(C_{df}(t))_0$ are compared in Fig. 10.5. It is clear that the curve identification by ILC is much more robust in initial curve determination than the existing methods in [44, 52].

Case 4: Identification Results with Learning Parameters Varying w.r.t. Learning Iteration Number

In **Case 1**, high-order and time-varying schemes (10.30), (10.31) have been shown to be very effective. When considering the schemes (10.32) and

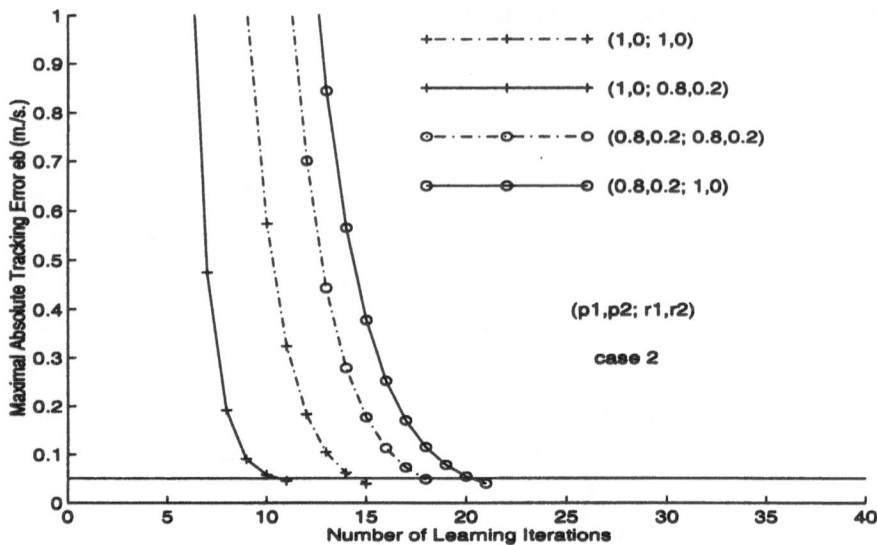

Fig. 10.4. Convergence history comparison of tracking error bound e_b (**Case 2**)

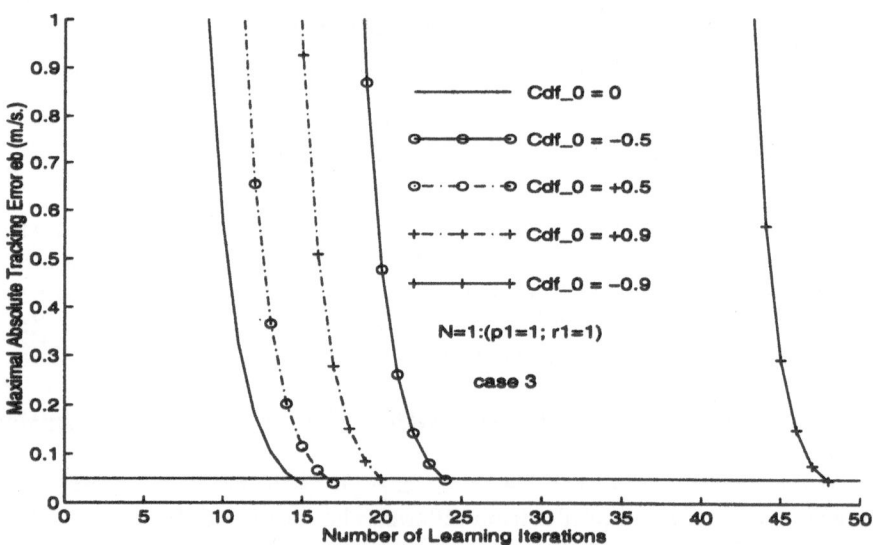

Fig. 10.5. Convergence history comparison of tracking error bound e_b (**Case 3**)

(10.33) in which the learning parameters vary w.r.t. both iteration number and time, even better results have been obtained as shown in Fig. 10.6. In this case, for a fair comparison, only first-order schemes are used. Similar to Fig. 10.3, two curves represent the constant learning parameter situations where $r_v = -0.015$ and $r_v = -0.01$. When $r_v = -0.02$, ILC will not converge. Thus the constant choices of learning parameters are not practical. When (10.32) and (10.33) are applied, much faster convergence has been achieved, which is clearly shown in Fig. 10.6.

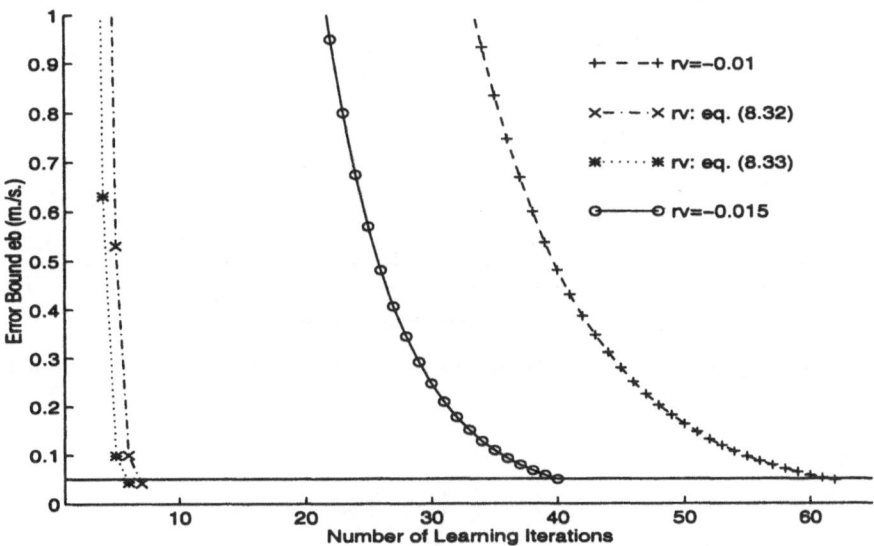

Fig. 10.6. Convergence history comparison of tracking error bound e_b (Case 4)

Case 5: Identification Results with Bi-linear Learning Scheme
The following three subcases have been tried.

- **Case 5.1.** *Constant r_v. $r_v = -0.015$. $\beta = 0, -0.01, -0.015$ respectively.*

  See Fig. 10.7 for e_b convergence history comparison. The number of ILC iterations required are 39, 19 and 19 respectively. Note the non-monotonic behavior of e_b when $\beta = -0.015$.
- **Case 5.2.** *Constant r_v. $r_v = -0.01$. $\beta = 0, -0.01, -0.02, -0.03$ respectively.*
  Fig. 10.8 shows the comparison of ILC convergence histories for e_b. When $\beta = 0, -0.01, -0.02$, the required iteration numbers are 61, 25, and 20 respectively. However, when $\beta = -0.03$, the iteration diverges.
- **Case 5.3.** *Varying r_v (scheme (10.32)). $\beta = 0, -0.00485$ respectively. With Comparison to the varying r_v (scheme (10.33)) when $\beta = 0$.*
  This subcase applies varying r_v schemes (10.32) and (10.33). In scheme (10.32), a bi-linear term is added ($\beta \neq 0$) which is shown in Fig. 10.9 in

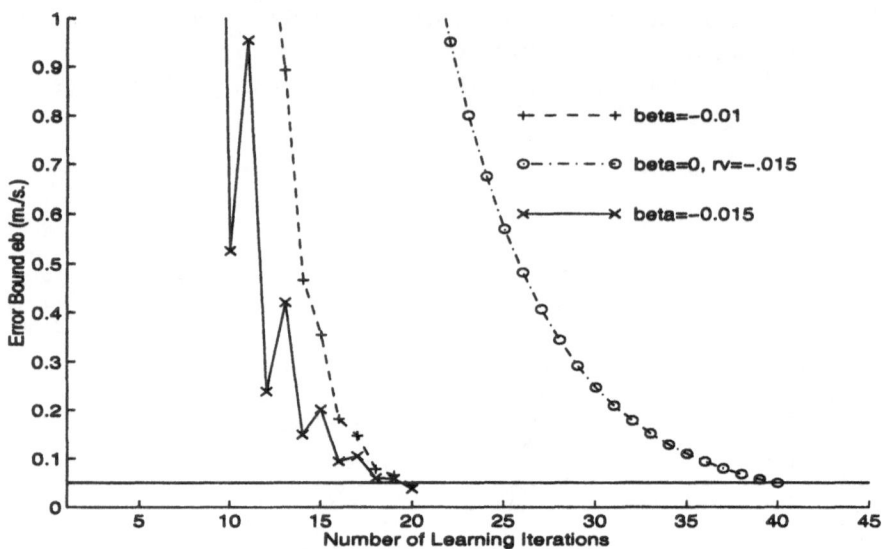

Fig. 10.7. Convergence history comparison of tracking error bound e_b (**Case 5.1**)

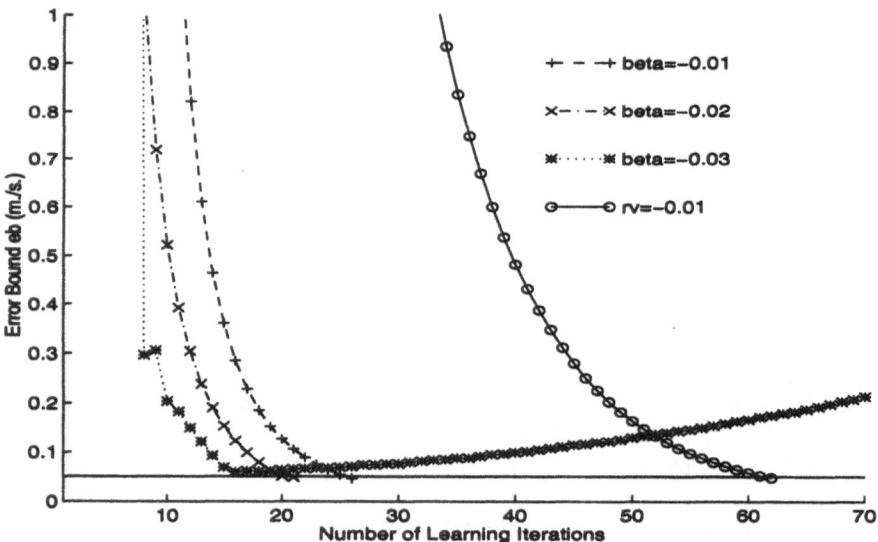

Fig. 10.8. Convergence history comparison of tracking error bound e_b (**Case 5.2**)

solid-line. Fig. 10.9 indicates that bilinear term is even beneficial to the optimally chosen r_v and only 4 learning iterations are actually required.

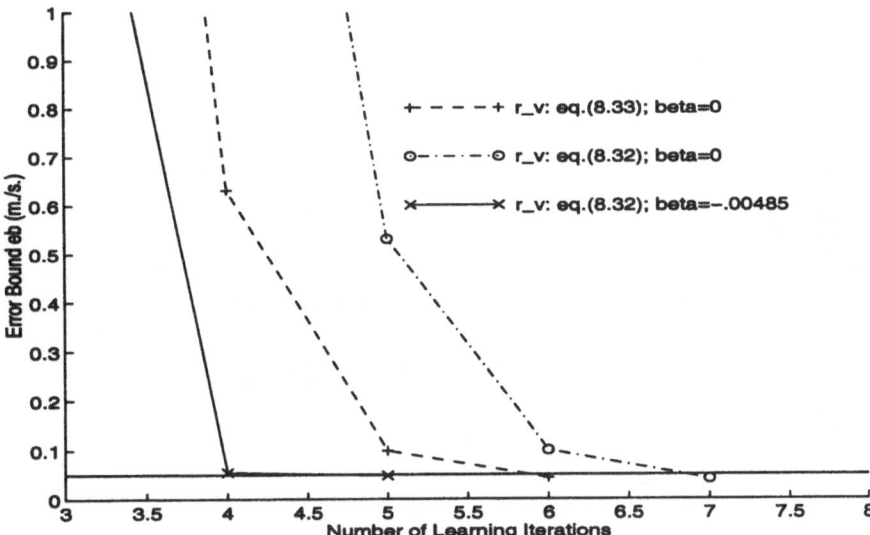

Fig. 10.9. Convergence history comparison of tracking error bound e_b (**Case 5.3**)

It should be noted that the converged drag coefficient curve must be a positive function. A comparison of the converged curves $C_{df}(M)$ between the optimal dynamic fitting method in [52] and the iterative learning identification is given in Fig. 10.10. From Fig. 10.10, the average value of the drag coefficient is about 0.28. Even though there are some negative initial curves $(C_{df}(t))_0$ in **Case 3**, the iterative learning identification still works well as shown in Fig. 10.5. Furthermore, the final converged results of this chapter with various learning parameter selections (**Case 1** to **Case 5**) are all the same and they are consistent with the result of optimal dynamic fitting method [52].

Based on the above practice of iterative learning identification of an unknown nonlinear function in a nonlinear dynamic system from testing measurements, some observations can be made as follows:

- Iterative learning identification is a very straightforward method and is simple to be implemented. The computation cost is low.
- The asymptotic convergence condition can be guaranteed and is easy to be satisfied. A suitable choice of learning parameters are easy to make, especially for the aerodynamic identification problem of this chapter.
- High-order learning scheme may supply much better performance over the first-order one. This is well illustrated by Fig. 10.3 and Fig. 10.4.
- Learning parameter varying with respect to time and iteration number gives even better learning performance as shown in Fig. 10.4 and Fig. 10.6.

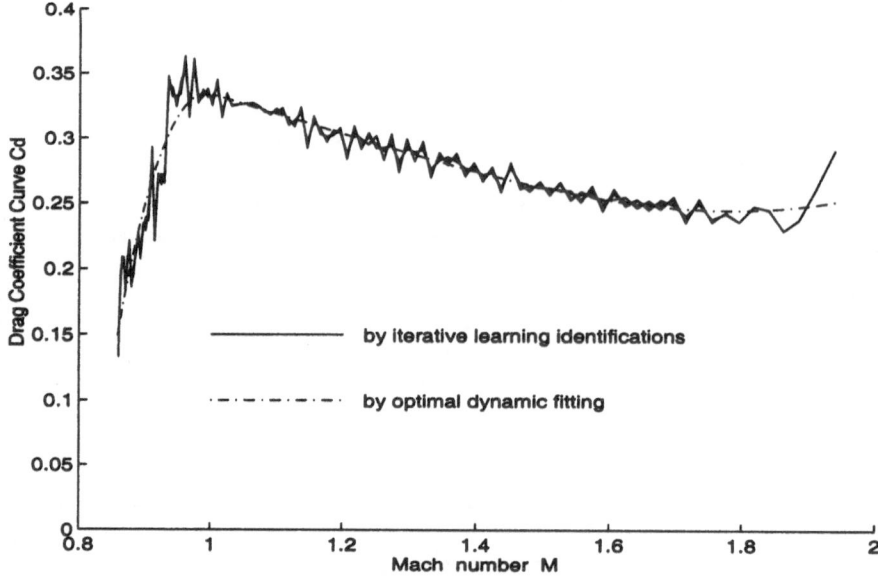

Fig. 10.10. Comparison of the identified $C_{df}(M)$'s

- A simple nonlinear learning updating law – bi-linear updating formula, has been suggested and tried. Some attractive results have been obtained. See Fig. 10.9. However, the learning convergence behavior become more complex and requires further investigations.
- The convergence of iterative learning identification is global. That is, the convergence is not sensitive to initial curve trials. This is clearly demonstrated in Fig. 10.5
- The performance index is in a form equivalent to the minimax one which is more favorable in engineering practice.

10.8 Conclusion

The iterative learning identification method is proposed for effectively extracting projectile's optimal fitting drag coefficient curve C_{df} from radar measured velocity data. The performance index is in a *minimax* sense and the initial curve can be arbitrarily chosen. A general convergence analysis is given. A guideline for choosing learning parameters, varying w.r.t. both time t and iteration number i, is suggested. A bi-linear scheme is proposed for an even better convergence speed. Its related identification results are also given. From the results of this chapter, it is clearly demonstrated that the iterative learning idea is very effective in identifying a nonlinear function in a nonlinear dynamic system.

11. Iterative Learning Control of Functional Neuromuscular Stimulation Systems

11.1 Motivations

BEING PARALYZED is sometimes believed to be worse than contracting cancer because it may change the life style of the whole family for a life-long time. To improve the life quality of people with neurological disorders caused by spinal cord injury, head injury and stroke etc., a rehabilitative technology using Functional Neuromuscular Stimulation (FNS) has been developed with increasing progresses in recent years. To rebuild or to restore the functions of existing muscles which are out of control of the central nerve system (CNS), computer controlled electrical stimuli are applied to the paralyzed muscles via electrodes. One of the key problems in FNS system is the *decision and control strategy* (DCS) which produces the required stimulation parameters of the stimulator to make the limb follow the pre-planned trajectory complete a desired task. In the DCS, three problems must be considered before the FNS system can be used on a clinical basis, i.e.,

- *customization problem* due to physiological differences between different paralyzed people,
- *adaptation problem* raised from the time-varying nonlinear musculoskeletal system properties
- *robustness problem* for the exogenous disturbances.

The existing DCS's [223, 178] did not address the above three problems at a whole extent until a neural network (NN) method [1] was introduced. This scheme is actually a combination of an adaptive feedforward controller and a fixed-parameter PD feedback controller. The adaptive feedforward controllers plus a feedback controller is said to be the *representative of the diverse control approaches that are likely to be required in future neuroprosthesis* [69].

There are many schemes to construct the feedforward controller. For example, a neural network is applied in [1]. Other machine intelligence based strategies like adaptive logic network, fuzzy logic etc. are also taken into considerations in [69]. In [75, 76], another new scheme termed *Iterative Learning Control* was proposed for the robust feedforward control of FNS systems. This chapter aims to provide a detailed discussion on the effectiveness of ILC applied to FNS systems.

Refer to Fig. 11.7. A musculoskeletal model which consists a single skeletal segment actuated by an agonist-antagonist pair of electrically stimulated muscles is used for the simulation studies. The musculoskeletal model is detailed in the next section. The control objective is to find a desired pulse width (PW) history $Z_j(t)$ $(j = 1, 2)$ (1: flexor; 2:extensor) to stimulate the muscle pair via the PWM stimulator in order to drive the skeletal segment to track the desired joint angle $\theta_d(t)$ and angular velocity $\dot{\theta}_d(t)$ which are planned to perform a given locomotive task. Note that $Z_1(t)$ and $Z_2(t)$ are not independent The iterative learning controllers discussed in Chapter 3 and Chapter 5 are simple to implement. But to a specified system, there are always some practical and simplified ways.

11.2 A Musculoskeletal Model

As the main purpose is to demonstrate another new application field of ILC, here a single skeletal segment is assumed for simplicity. The segment is driven by an agonist-antagonist pair of muscles across the joint as shown in Fig. 11.7. The skeletal dynamics with linear stiffness and damping acting to resist the movement around the joint is described as [1]

$$\sum_{j=1}^{2} T_{m_j}(t) + T_n(t) = (\frac{1}{4}ml^2 + \frac{1}{2}I)\ddot{\theta}(t) + \frac{1}{2}mgl \sin \theta(t)$$

$$+K_S(\theta(t) - \theta_0) + K_B\dot{\theta}(t) \qquad (11.1)$$

where T_{m_j} $(j = 1, 2)$ are the muscle torques generated by flexor and extensor muscles; T_n is the disturbance torque; m, l, I are mass, length, moment of inertia of the skeletal segment respectively; g is the gravitational acceleration; K_S, K_B are joint stiffness constant and damping constant respectively; θ_0 is the joint stiffness reference angle. The iterative learning repetition number i is dropped in (11.1). The joint stiffness reference angle θ_0 is normally set to 0 if the desired trajectory $\theta_d(0) = 0$. The model parameters in (11.1) are listed in Table 11.1.

Table 11.1. Skeletal Model Parameters

para.	m	I	l	K_S	K_B	g
values	10	0.1	0.4	20	1.0	9.80
units	kg	kgm^2	m	Nm/rad	Nm/°/s	m/s^2

Nonlinear muscle recruitment, linear muscle dynamics in force generation, and multiplicative nonlinear torque-angle and torque-velocity factors are considered in the muscle torque generation model as shown in Fig. 11.1 [1].

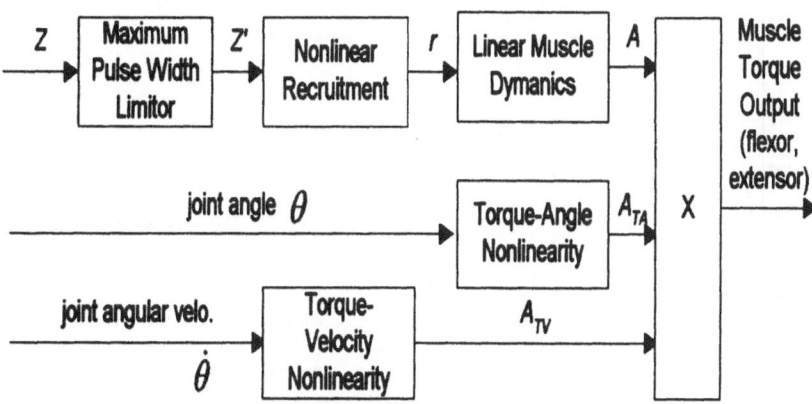

Fig. 11.1. Nonlinear Muscle Torque Generation Model

From Fig. 11.1, the muscle torque can be expressed as

$$T_{m_j}(t) = A_j(t) A_{TA_j} A_{TV_j} \tag{11.2}$$

where $A_j(t), A_{TA_j} (\in [0, 1]), A_{TV_j} (\in [0, 1.8])$ are the muscle activation factor, torque-angle factor and torque-velocity factor respectively. The force/torque generation in the muscle satisfies the linear dynamic difference equation

$$A_j(t) = a_{1_j} A_j(t - t_s) + a_{2_j} A_j(t - 2t_s) + b_{0_j} r_j(t - t_s) \tag{11.3}$$

where t_s $(= 0.05s)$ is the fixed stimulation period (SP), r_j $(\in [0, 1])$ is the muscle recruitment, $a_1 (= 0.6679), a_2 (= -0.1001), b_{0_1} (= -b_{0_2} = 20Nm)$ are the coefficients of the linear dynamics which may be changing due to muscle fatigue as discussed later. Set the full muscle recruitment, i.e., $r_j = 1$, the linear dynamic response is shown in Fig. 11.2.

The nonlinear recruitment r_j of the flexor and extensor is given as

$$r_j(t) = \gamma_0 + \gamma_1 (Z' - D) + \gamma_2 (Z' - D)^2 + \gamma_3 (Z' - D)^3 \tag{11.4}$$

where $D(= 3.501\mu s)$ is the recruitment deadband;

$$[\gamma_0, \gamma_1, \gamma_2, \gamma_3] = [0, 0.01909, -0.0001152, 0.00000026]$$

where $\gamma_j (j = 0, \cdots, 3)$ are the recruitment curve coefficients; $z' \in [0, 100]\mu s$ is the output of the PW limiter as shown in Fig. 11.1. The unit of the PW is **microsecond** (μs). This is fully plotted in Fig. 11.3.

The torque-angle nonlinearity factor A_{TA_j} is

$$A_{TA_j}(t) = 1 - (\theta(t) - \bar{\theta}_{0_j})^2 / \theta_{w_j}^2 \tag{11.5}$$

where A_{TA_j} is dimensionless and clamped in $[0, 1]$, $\bar{\theta}_{0_1}$ $(= -\bar{\theta}_{0_2} = -20°)$ is the reference angle and θ_{w_j} $(= 100°)$ is the torque-angle width. The torque-velocity nonlinearity factor A_{TV_j} is expressed as

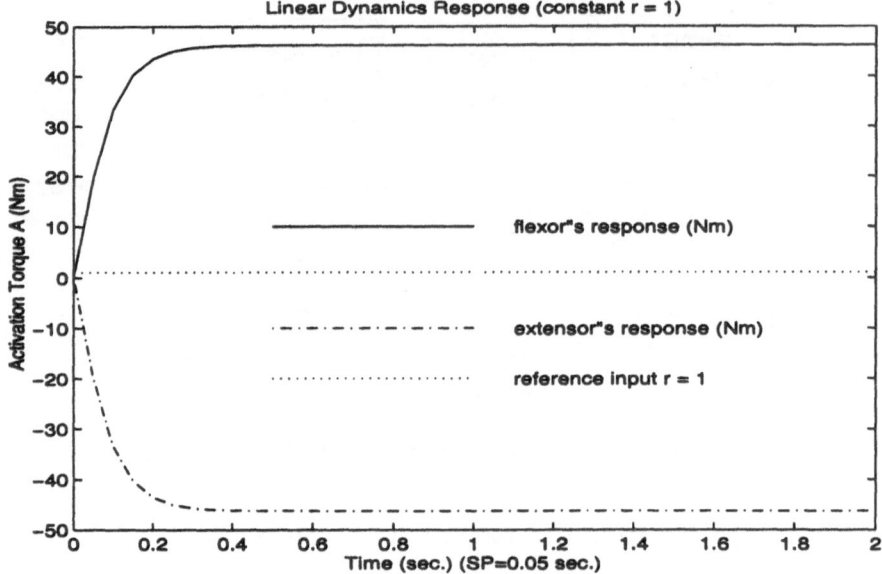

Fig. 11.2. Linear Dynamic Response of Muscle Torque Generation

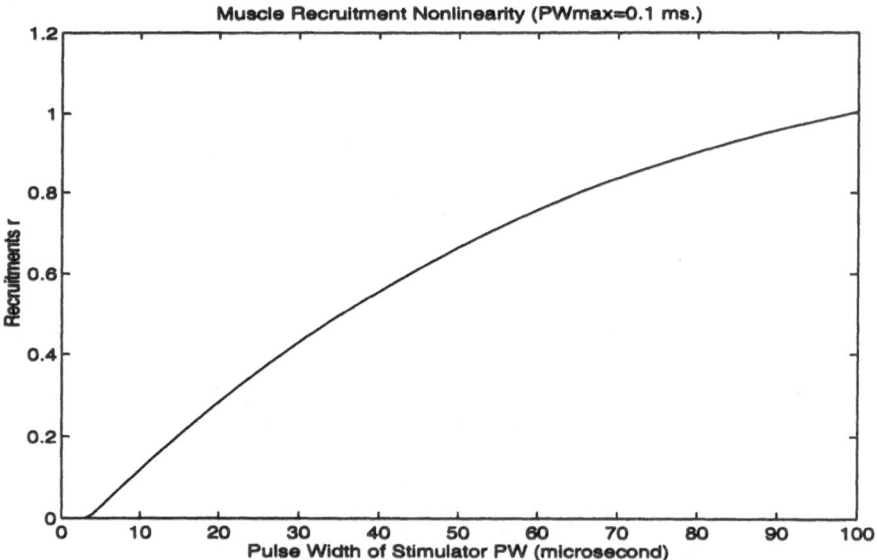

Fig. 11.3. Nonlinear Muscle Recruitment Curve

$$A_{TV_j}(t) = \frac{k_{\dot{\theta}_j}[\dot{\theta}_{max_j} - \dot{\theta}]}{k_{\dot{\theta}_j}\dot{\theta}_{max_j} + \dot{\theta}} \tag{11.6}$$

where A_{TV_j} is dimensionless and clamped in $[0, 1.8]$, $k_{\dot{\theta}}$ ($= 0.2$) is the torque-velocity slope, $\dot{\theta}_{max}$ ($= 600°/s$) is the maximum shortening velocity. These nonlinearities are shown in Figs. 11.4 and 11.5.

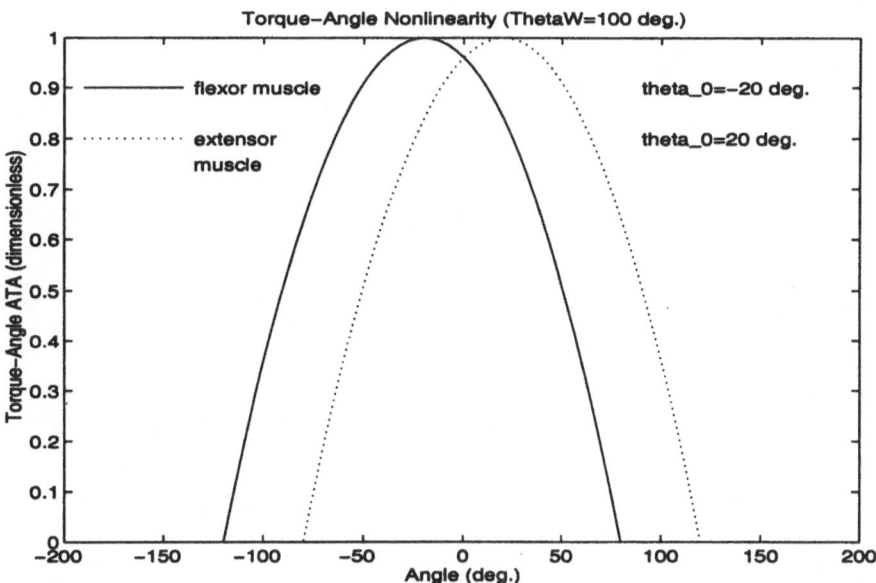

Fig. 11.4. Muscle Torque-angle Nonlinearity

The muscle fatigue was also included to further simulate the realistic situation of FNS characteristics such that the simulation results of this chapter can be actually directive for the FNS DCS investigation and determination. The muscle fatigue model consists of a gradually decreasing muscle's input gain (b_{0_j}) to 50% of its original value (20Nm) over 100 s

$$b_0(t) = b_0(t - t_s) + k_f[b_{0_f} - b_0(t - t_s)] \tag{11.7}$$

where b_{0_f} ($= 10$Nm) is the final value of $b_0(t)$, k_f ($= 0.002$) is a fatigue rate factor. Fig. 11.6 shows the muscle fatigue curves.

The musculoskeletal model described above is to simulate the realistic situation of FNS characteristics such that the simulation results can be actually useful for the FNS DCS study.

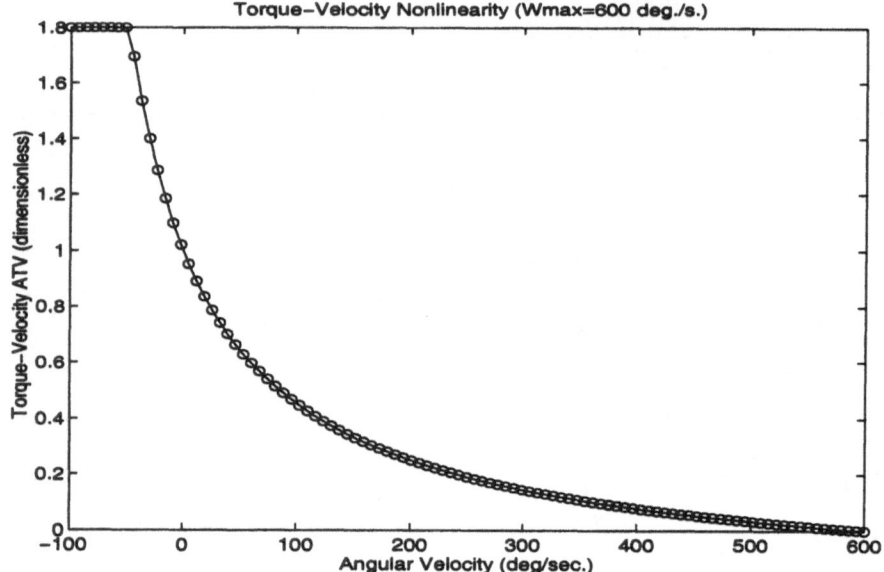

Fig. 11.5. Muscle Torque-velocity Nonlinearity

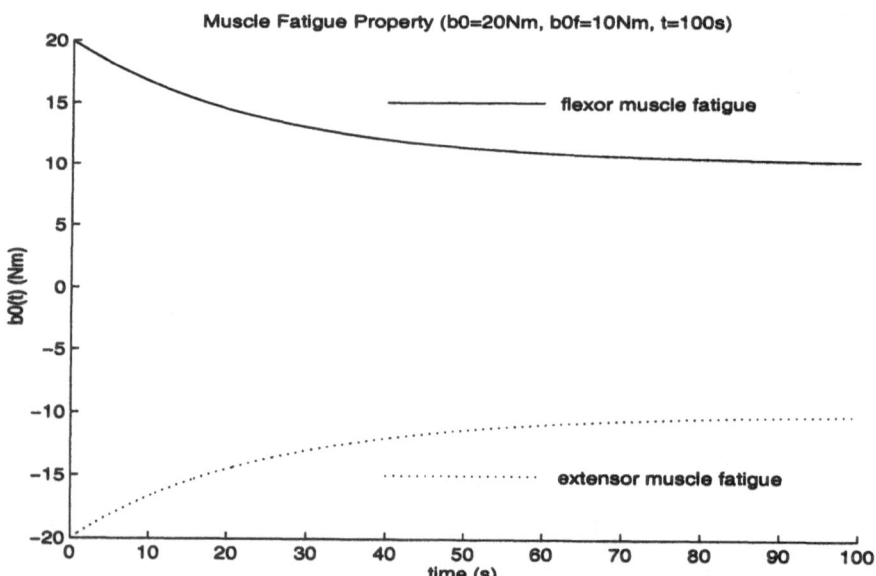

Fig. 11.6. Muscle Fatigue Curve

11.3 Simplified First-order Continuous-time ILC Scheme

As shown similarly in Fig. 11.7, the controller is in a quite simple mode which includes a PD-type feedback controller and a first-order iterative learning controller. The first-order ILC updating law is in a continuous-time form as follows.

$$\bar{Z}_j^{(i+1)}(t) = \bar{Z}_j^{(i)}(t) + \bar{k}_p e_\theta^{(i)}(t) + \bar{k}_d e_{\dot\theta}^{(i)}(t) \tag{11.8}$$

$$Z_j^{(i)}(t) = \bar{Z}_j^{(i)}(t) + k_p e_\theta^{(i)}(t) + k_d e_{\dot\theta}^{(i)}(t) \tag{11.9}$$

where i is the task repetition number, $\bar{k}_p, \bar{k}_d$ and k_p, k_d are the learning gains and the PD gains respectively, $e_\theta = \theta_d(t) - \theta(t)$ and $e_{\dot\theta} = \dot\theta_d - \dot\theta$ are tracking errors. In this situation, it is required that the joint angle and the joint angular velocity must be measured for the human limb movement. Because no derivative information of the state $[\theta(t), \dot\theta(t)]$ is required, this scheme is referred to as a P-type ILC algorithm.

It should be pointed out that, according to the result of Chapter 3, to have the control function appear in the output equation, the acceleration of joint angle must be used. However, the convergence result still holds as shown in the simulations. More rigorous discussions can be found in [132, 218]. This chapter focuses on the discrete-time form of ILC scheme for FNS systems.

11.4 Simplified Second-order Discrete-time ILC Scheme

A high-order P-type discrete-time ILC law ($M = 2$) combined with a conventional PD feedback controller is applied which is described as follows

$$\hat{Z}_j^{(i)}(t) = \hat{Z}_j^{(i-1)}(t) + \hat{k}_{1_p} e_\theta^{(i-1)}(t+1) + \hat{k}_{1_d} e_{\dot\theta}^{(i-1)}(t+1)$$
$$+ \hat{k}_{2_p} e_\theta^{(i-2)}(t+1) + \hat{k}_{2_d} e_{\dot\theta}^{(i-2)}(t+1) \tag{11.10}$$

$$Z_j^{(i)}(t) = \hat{Z}_j^{(i)}(t) + k_p e_\theta^{(i)}(t) + k_d e_{\dot\theta}^{(i)}(t) \tag{11.11}$$

where i is the task repetition number, $\hat{k}_{1_p}, \hat{k}_{1_d}, \hat{k}_{2_p}, \hat{k}_{2_d}$ and k_p, k_d are learning gains and PD gains respectively, $e_\theta = \theta_d(t) - \theta(t)$ and $e_{\dot\theta} = \dot\theta_d - \dot\theta$ are the tracking errors at discrete time instant t.

It should be pointed out here that, based on the above discrete-time ILC setting and the theoretical analysis in Chapter 5, it is not necessary to measure joint angular velocity as well as the acceleration. This is demonstrated by the simulation.

11.5 Simulation Results and Discussions

Based on the musculoskeletal model detailed in Sec. 11.2, a series of simulations are performed to validate the effectiveness of the proposed iterative learning control method for FNS systems. In the following cases, the desired tracking trajectories are specified as [242]

$$\theta_d(t) = k\{\theta_b + (\theta_b - \theta_f)(15\tau^4 - 6\tau^5 - 10\tau^3)\} \tag{11.12}$$

$$\dot{\theta}_d(t) = k(\theta_b - \theta_f)(60\tau^3 - 30\tau^4 - 30\tau^2) \tag{11.13}$$

where $\tau = t/(t_f - t_0)$ and the velocity profile is in a bell form. We choose $\theta_b = 0°$, $\theta_f = 10°$, $t_0 = 0$, $T \stackrel{\triangle}{=} t_f = 1.0\ s$. In some cases, we choose $\theta_f = 15°$ and $\theta_f = 20°$ to show the nonlinear behavior of the muscle model. Fig. 11.8 shows $\theta_d(t)$ and $\dot{\theta}_d(t)$ for the cases $\theta_f = 10°$, $15°$, $20°$ with all $\theta_b = 0°$. It is worthwhile to note that $\dot{\theta}_d(0) = \dot{\theta}_d(T) = 0$ and $\ddot{\theta}_d(0) = \ddot{\theta}_d(T) = 0$ which represent the natural movement of human limb [242].

11.5.1 Fundamental Tests

To validate the model in Sec. 11.2 can generate a reasonable muscle torque, several fundamental tests are to be carried out. The tests are also helpful in planning a reasonable desired trajectory, i.e., θ_f such that the required muscle torque is within what can be generated.

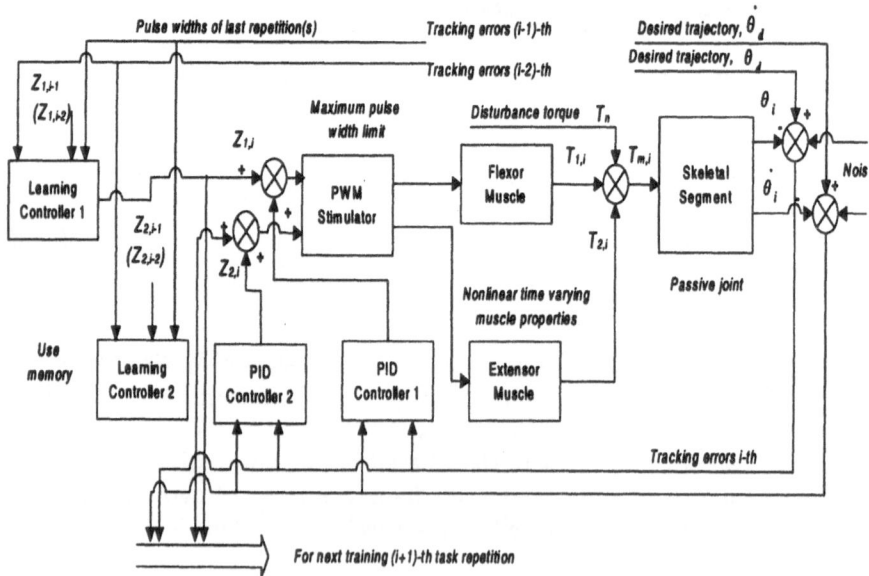

Fig. 11.7. The High-order ILC Strategy for FNS ($M = 2$)

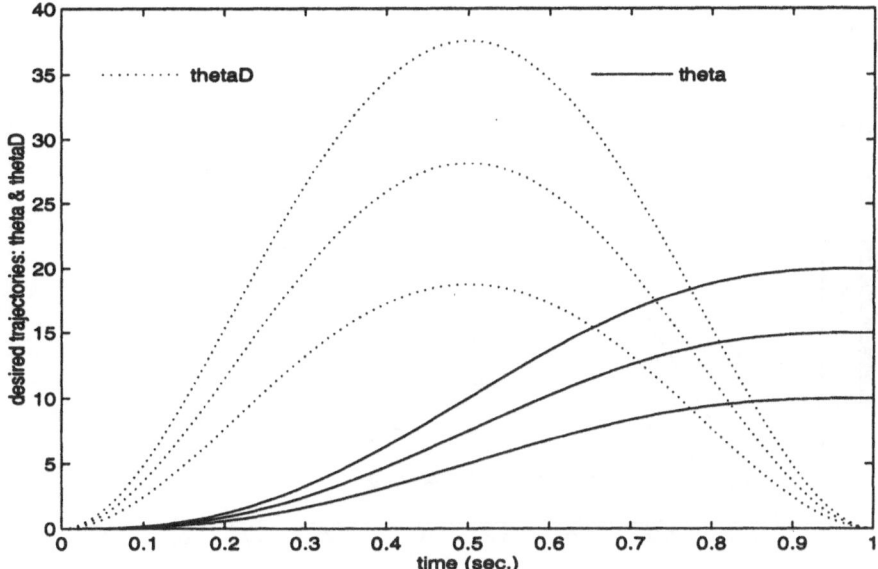

Fig. 11.8. The desired trajectories: joint angle and angular velo.

Direct Joint Torque Driven. As shown in Fig. 11.2 in Sec. 11.2, the maximum muscle activation torque A is approximately ±47 Nm. We first examine the desired muscle torque required for the desired trajectory by assuming the the direct joint torque driven. The major objective for investigating the desired $T_{m_j}(t)$ in (11.1) under different θ_f's is to validate the muscle torque generation model and to determine a suitable desired θ_f for later simulation sessions.

In this situation, the ILC updating law we used is

$$(T_m(t))_{i+1} = (T_m(t))_i + \gamma_p(e_\theta)_i + \gamma_v(e_{\dot\theta})_i$$

where γ_p and γ_v are chosen as 1.5 Nm/rad. and 5 Nm/rad./s. respectively. Note that these two learning gains are to guarantee the learning convergence and not necessarily to be 'optimal'. Set the exit accuracy of ILC iterations as

$$e_{b_1} \stackrel{\triangle}{=} \sup_{t \in [0,1]} \mid (e_\theta) \mid \leq 2° \text{ and } e_{b_2} \stackrel{\triangle}{=} \sup_{t \in [0,1]} \mid (e_{\dot\theta}) \mid \leq 2°/s.$$

Set all initial conditions to 0. Then, for the desired trajectories of $\theta_f = $ 10°, 15°, 20°, the required numbers of ILC iterations are 29, 34, 37 respectively. The final *iteratively learned* desired muscle torques $T_m(t)$ are given in Fig. 11.9. It is clear from these results that, according to the maximal muscle torque that can be generated as shown in Sec. 11.2, a suitable θ_f should be around 15°.

The ILC convergence processes are shown in Fig. 11.10.

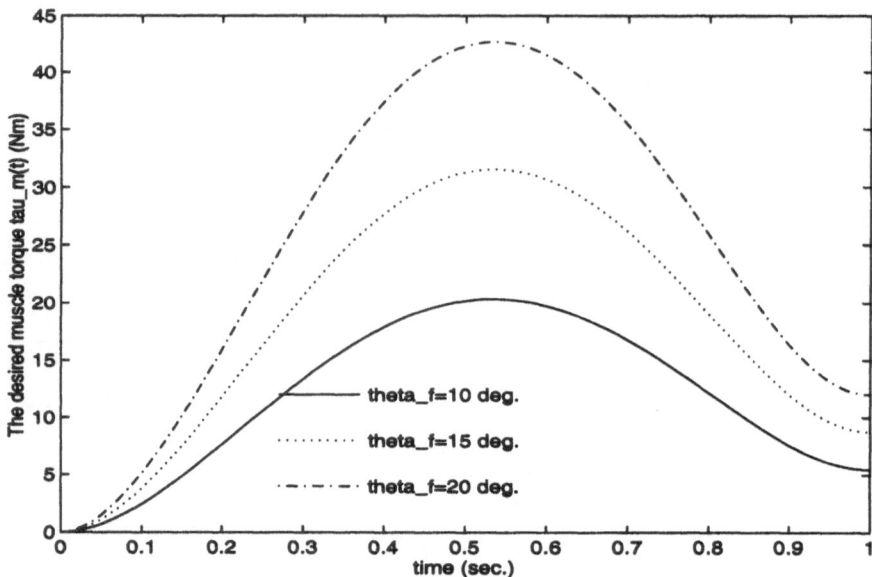

Fig. 11.9. The *iteratively learned* desired muscle torques

Conventional PD-controller. A PD closed-loop conventional controller alone is applied with fixed parameters $k_p = 10000$ μs/rad, $k_d = 500$ μs/rad/s [1]. The PD parameters may not be well tuned and they are just used for illustrations. The muscle model given in Sec. 11.2 is used without considerations of the muscle fatigue and the disturbance.

Two cases are considered.

- Case 1: Flexor activation only. ($Z_2(t) \equiv 0$)
  Referring to Fig. 11.9, the desired inputs for the desired trajectories (11.13) are positive. It is natural to consider the usage of only one muscle for the muscle activation. In this case, only flexor activation is sufficient. Figs. 11.15 - 11.17 show the controls and control performances at different θ_f. In the top graph of Figs. 11.15, y_1 and y_{d_1} represent θ and θ_d respectively; y_2 and y_{d_2} represent $\dot{\theta}$ and $\dot{\theta}_d$ respectively. This abbrevation is valid in all figures if not otherwise indicated. The nonlinear effects of the FNS system model to the the fixed parameter conventional PD controller suggest that an advanced control strategy is in demand.
- Case 2: Two muscle activation.
  Now consider the case of two muscle activation. When in actual movement, there is a time duration when the flexor and the extensor are activated simultaneously but with different level of activation. This is called the *co-activation* and implies that for the stimulator of the FNS system, the stimuli should be assigned to the flexor or the extensor through a *co-*

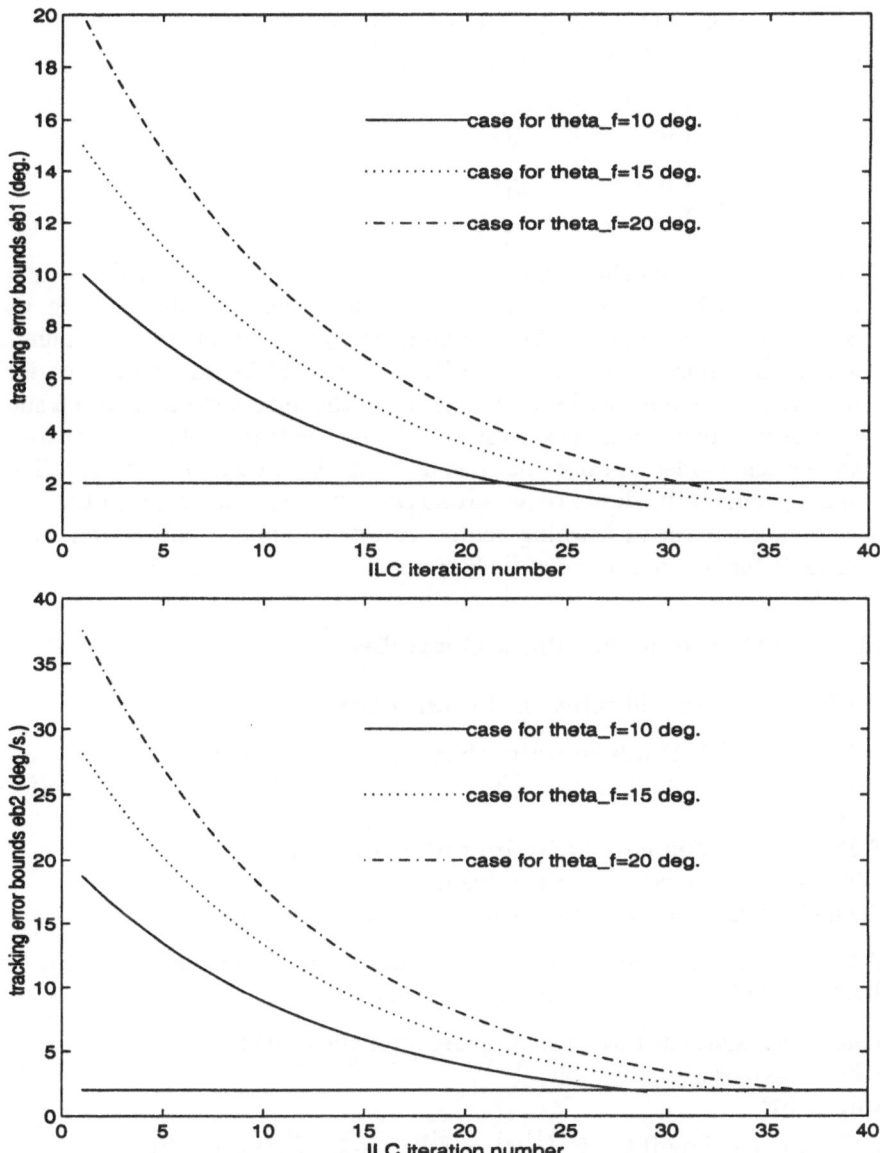

Fig. 11.10. Test: ILC convergence processes, joint angle and joint angular velocity

contraction map [223, Chapter 14]. In this study, to make the simulation process clear, the following simple *'map'* is used:

$$Z_1(t) = \begin{cases} Z_1(t), & \text{if } Z_1(t) \geq 0; \\ 0, & \text{if } Z_1(t) < 0; \end{cases}$$

$$Z_2(t) = \begin{cases} 0, & \text{if } Z_1(t) \geq 0; \\ -Z_1(t), & \text{if } Z_1(t) < 0; \end{cases}$$

Figs. 11.18 - 11.20 show the controls and control performances at different θ_f under flexor-extensor muscle activation. Conceptually speaking, the control performances should be improved by the flexor-extensor muscle activation. However, compared to Figs. 11.15 - 11.17, the control performances are not improved significantly from the introduction of such a simple flexor-extensor muscle activation. The reason is the nonlinear nature of the muscle model and notable time-delay in the muscle recruitment. This again prompts the desire of an advanced control for such case. In the next section, an iterative learning controller will be applied and shown to be suitable for better control of FNS systems.

11.5.2 ILC without Feedback Controller

In this section, we will verify the following facts:

Fa The discrete ILC will be better than a continuous version ILC.
Fb The high-order ILC updating law gives a better ILC convergence performance.
Fc The ILC method is still effective with muscle fatigue
Fd The ILC can reject the repeatable uncertainty and disturbances.
Fe The ILC can track on the slowly varying desired trajectory.

ILC Implementation: Continuous Form versus Discrete Form. To illustrate [Fa], a simple scenario is set as follows:

- no uncertainty/disturbance, no muscle fatigue considered,
- flexor-extensor muscle pair,
- $\theta_f = 10°$,
- Continuous-time ILC: $Z_j^{(i+1)}(t) = Z_j^{(i)}(t) + \bar{k}_p e_\theta^{(i)}(t) + \bar{k}_d e_{\dot{\theta}}^{(i)}(t)$,
- Discrete-time ILC: $Z_j^{(i+1)}(t) = Z_j^{(i)}(t) + \bar{k}_p e_\theta^{(i)}(t+1) + \bar{k}_d e_{\dot{\theta}}^{(i)}(t+1)$,
- $\bar{k}_p = 100\mu s/\text{rad}$, $\bar{k}_d = 50\mu s/\text{rad/s}$,
- full 12 ILC iterations performed.

The ILC convergence processes are compared in Fig. 11.11. It can be seen that a discrete implementation of continuous ILC will be slightly better than the continuous ILC updating formula. The advantage is from the $(t+1)$ which contains derivative information as discussed in Remark 5.2.1.

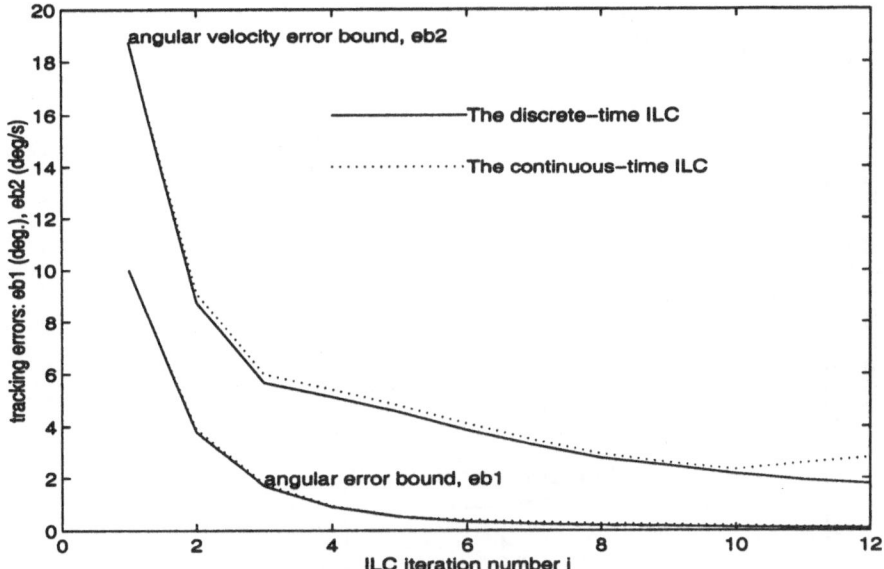

Fig. 11.11. ILC convergence comparison, discrete vs. continuous

High-order ILC. Consider the same case of Sec. 11.5.2, but the ILC updating law is in a second-order discrete-time form

$$Z_j^{(i+1)}(t) = Z_j^{(i)}(t) + \bar{k}_{p_1} e_\theta^{(i)}(t+1) + \bar{k}_{d_1} e_{\dot\theta}^{(i)}(t+1)$$
$$+ \bar{k}_{p_2} e_\theta^{(i)}(t+1) + \bar{k}_{d_2} e_{\dot\theta}^{(i)}(t+1)$$

where $\bar{k}_{p_1} = 100\mu s/rad$, $\bar{k}_{d_1} = 50\mu s/rad/s$ and $\bar{k}_{p_2} = 0.1\bar{k}_{p_1}$, $\bar{k}_{d_2} = 0.1\bar{k}_{d_1}$. Fig. 11.12 clearly demonstrates [Fc].

To have a comparison of the controls and the control performances, a figure similar to Figs. 11.18 - 11.20 and Figs. 11.15 - 11.17 for the high-order ILC case is given in Fig. 11.21, which is much better than the conventional PD-controller.

High-order ILC with Muscle Fatigue. By using the same setups of Sec. 11.5.2, the muscle fatigue effects are illustrated. The ILC updating is in a second-order form and the parameters are the same as those used in Sec. 11.5.2. Because only 12 ILC iteration are performed, the FNS system lasts total 12 seconds. The gain $b_0(t)$ of muscle linear dynamics varies from 20 Nm to 13.8 Nm for flexor and -20 Nm to -13.8 Nm for extensor respectively.

In Fig. 11.13, the ILC convergence process is compared with the one without muscle fatigue as obtained in Sec. 11.5.2 for the case of using second-order discrete-time ILC updating law. Fig. 11.22 shows the control function and the tracking performance of the ILC affected by muscle fatigue. For a clear

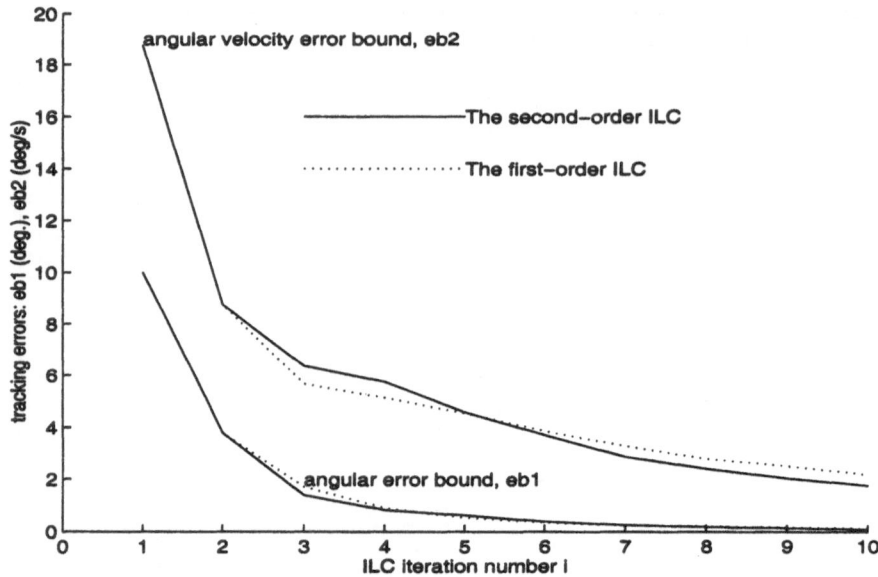

Fig. 11.12. ILC convergence comparison, the first-order vs. the second-order

comparison, the relevant curves from the case without considering muscle fatigue effect are also plotted in Fig. 11.22. It can be observed that the muscle fatigue will deteriorate the control performance. However, in this study, we can see that the muscle torque generated is gradually less than the required one which means that the system is beyond its control capacity. If the muscle fatigue reduces the muscle torque slowly and especially, if the magnitude of the torque makes the desired trajectory feasible, the ILC method of this chapter can overcome this time-varying gain problem as assured in the theoretical investigations, see Chapter 6. In this simulation, the fatigue reduces the muscle torque to an infeasible region which produces the *steady state* tracking error as shown in Fig. 11.22.

High-order ILC under Exogenous Torque Disturbance. The simulation settings are the same as Sec. 11.5.2. To examine the robustness of ILC to disturbance, an exogenous mechanical torque disturbance is considered. Assume the disturbance is time-varying but invariant with respect to ILC iteration number i which is called the *repetitive disturbance*. The disturbance $T_n(t)$ in (11.1) is set as follows:

$$T_n(t) = 10\sin(8\pi t/N) \quad \text{Nm}, \ t = 1, 2, \cdots, N.$$

The ILC convergence is shown in Fig. 11.14. The control function and the tracking performance are shown in Fig. 11.23. It clearly verifies the theoretical results of Sec. 5.3.2 which proved that the repetitive disturbance can be rejected from the tracking error bound. It should be pointed out that if the disturbance or the uncertainty is not repeatable, the tracking error bound is a function of the bound of the difference of the disturbances between two successive ILC iterations. This can also be verified easily and is omitted here. To this point, [Fd] has been verified.

High-order ILC for Varying Desired Trajectories. The simulation settings are the same as Sec. 11.5.2. As discussed in Remark 5.3.5, the desired trajectory can be varying with respect to the ILC iteration number i under the restriction that the difference of the desired trajectories between two successive ILC iterations is bounded. This difference bound directly affects the final tracking error bound. In this study, the varying desired trajectory is simply set as

$$(\theta_f)_i = 5° + (i-1)/5.$$

Perform the ILC operation 50 times (No. 1 to No. 51) without consideration of muscle fatigue and disturbance. To show the effectiveness, an ILC tracking result for the fixed desired trajectory of $\theta_f \equiv 15°$ is used for comparison. As shown in Fig. 11.27, the ILC performance in the varying desired trajectory case is even better than the fixed case. This is a quite interesting phenomenon requiring future investigations. Generally speaking, this is a nonlinear behavior. The control function and the tracking performance are shown in Fig. 11.24, which validates [Fe].

11.5.3 ILC Plus a Feedback Controller

As argued in [102, 16], a feedback controller is used to stabilize the system. With this, the system output at the first ILC trial is much closer to the desired one compared to the zero output in the conventional open-loop ILC. Intuitively, a better ILC performance can be achieved with the help of a feedback controller. As the robustness issues for the open-loop ILC have been well demonstrated in Sec. 11.5.2, this section is just to show the improvements by the introduction of a feedback controller. The improvement is mainly in the ILC transients.

Three cases are considered:

- (Case 1), PD closed-loop conventional controller only ($k_p = 10000$ $\mu s/rad$, $k_d = 500$ $\mu s/rad/s$);
- (Case 2), ILC controller plus PD feedback controller, $M = 1$, with $\hat{k}_{1_p} = 500$ $\mu s/rad$, $\hat{k}_{1_d} = 100$ $\mu s/rad/s$;
- (Case 3), ILC controller plus PD feedback controller, $M = 2$, with $\hat{k}_{2_p} = -\frac{1}{8}\hat{k}_{1_p}$ $\mu s/rad$, $\hat{k}_{2_d} = -\frac{1}{8}\hat{k}_{1_d}$ $\mu s/rad/s$;

The histories of maximal absolute tracking errors of angular position and velocity obtained for Cases 2 and 3 are shown in Fig. 11.25. It can be observed that the high-order ILC gives better ILC convergence. Fig. 11.26 is the angular position and velocity tracking comparisons for all the three cases.

It is clear from these simulation results that *learning from repetitions* DCS does give a better control performance than conventional control methods. High-order ILC can improve ILC performance further. From the robustness property of ILC as analyzed in the previous chapters, it is also obvious that FNS system based on ILC DCS will possess better robustness which can cover the aforementioned three problems, i.e., the *customization, adaptation* and *robustness*, in practical FNS systems.

11.6 Conclusions

A high-order Iterative Learning Control strategy is proposed for a better control of FNS system. Robust convergence properties have been investigated for both continuous-time and discrete-time high-order P-type ILC schemes. A suitable musculoskeletal model is utilized to conduct the simulation studies. Through extensive simulations, it is shown that the ILC method can give a better solution to the *customization, adaptation* and *robustness* problems in FNS system control. Moreover, it should be pointed out that unlike the conventional adaptive control method, the adaptation via ILC method is in a *point-wise* fashion. The ILC DCS is hopeful to be suitable for the real task execution both in the training stage use and in the daily-life use of the FNS systems.

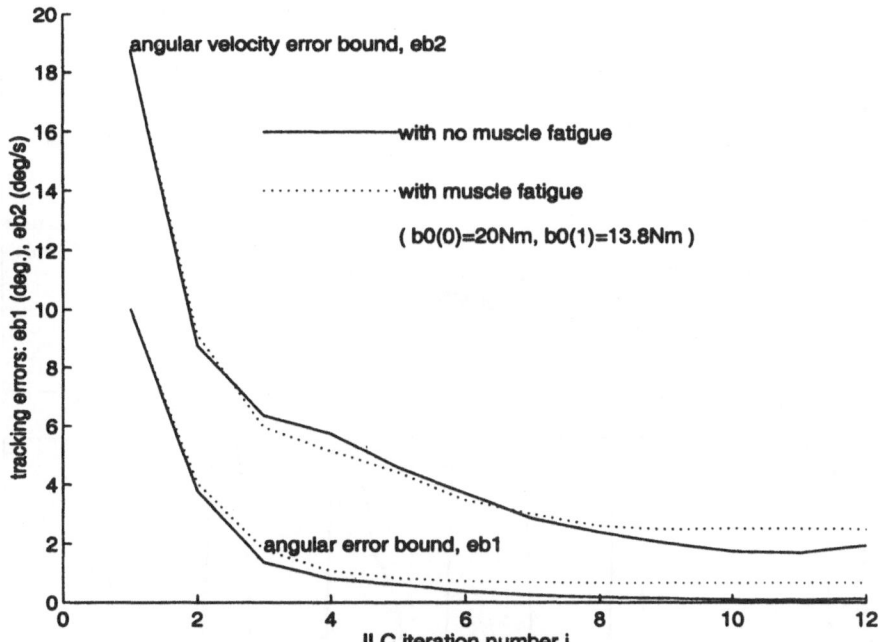

Fig. 11.13. ILC convergence comparison, muscle fatigue effect

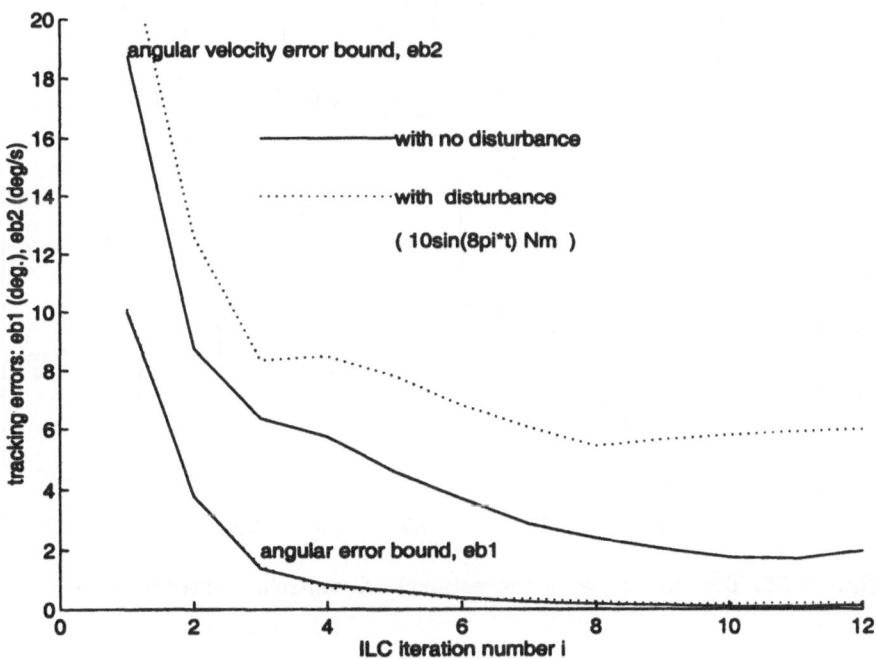

Fig. 11.14. ILC convergence comparison, exogenous torque disturbance rejection

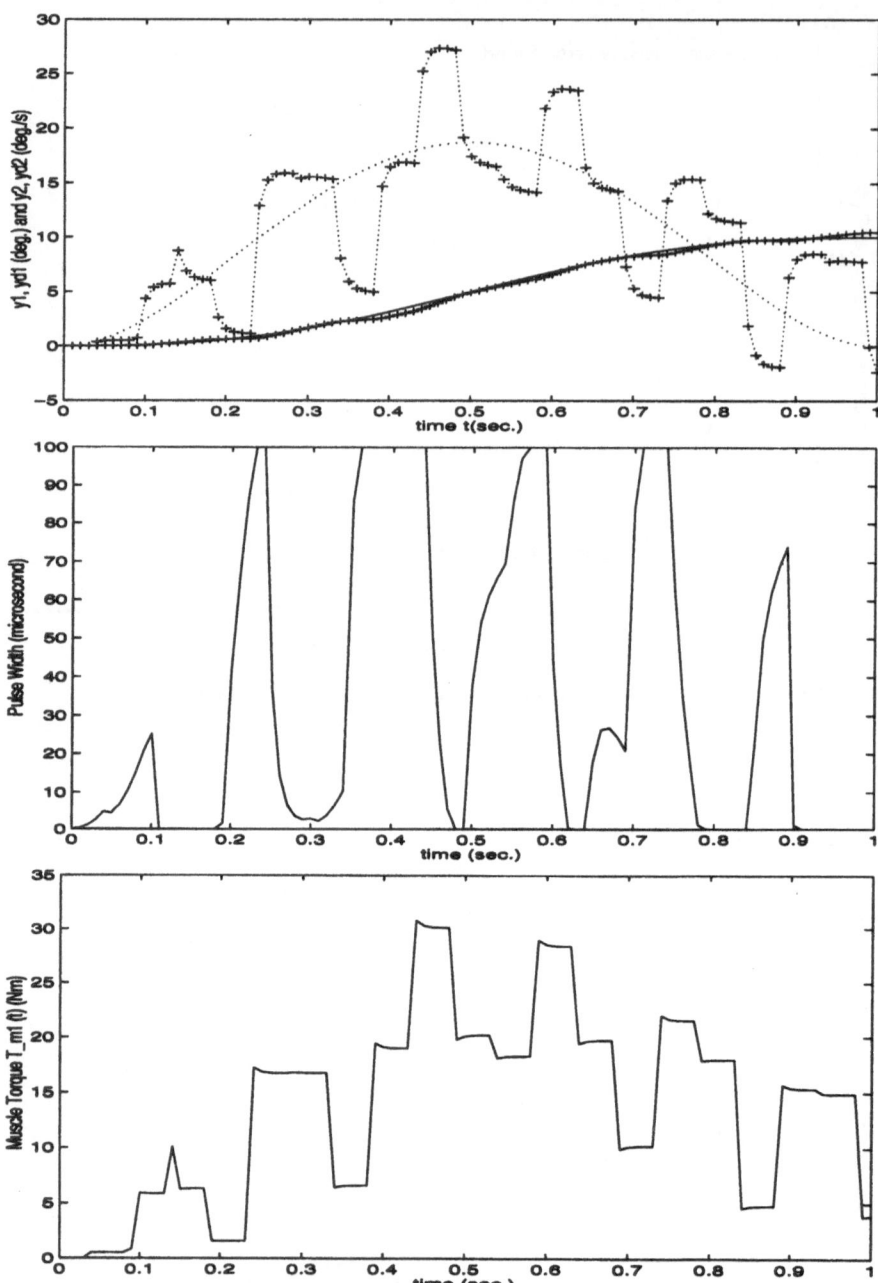

Fig. 11.15. Test: Responses of conventional PD controller, flexor only, $\theta_f = 10°$

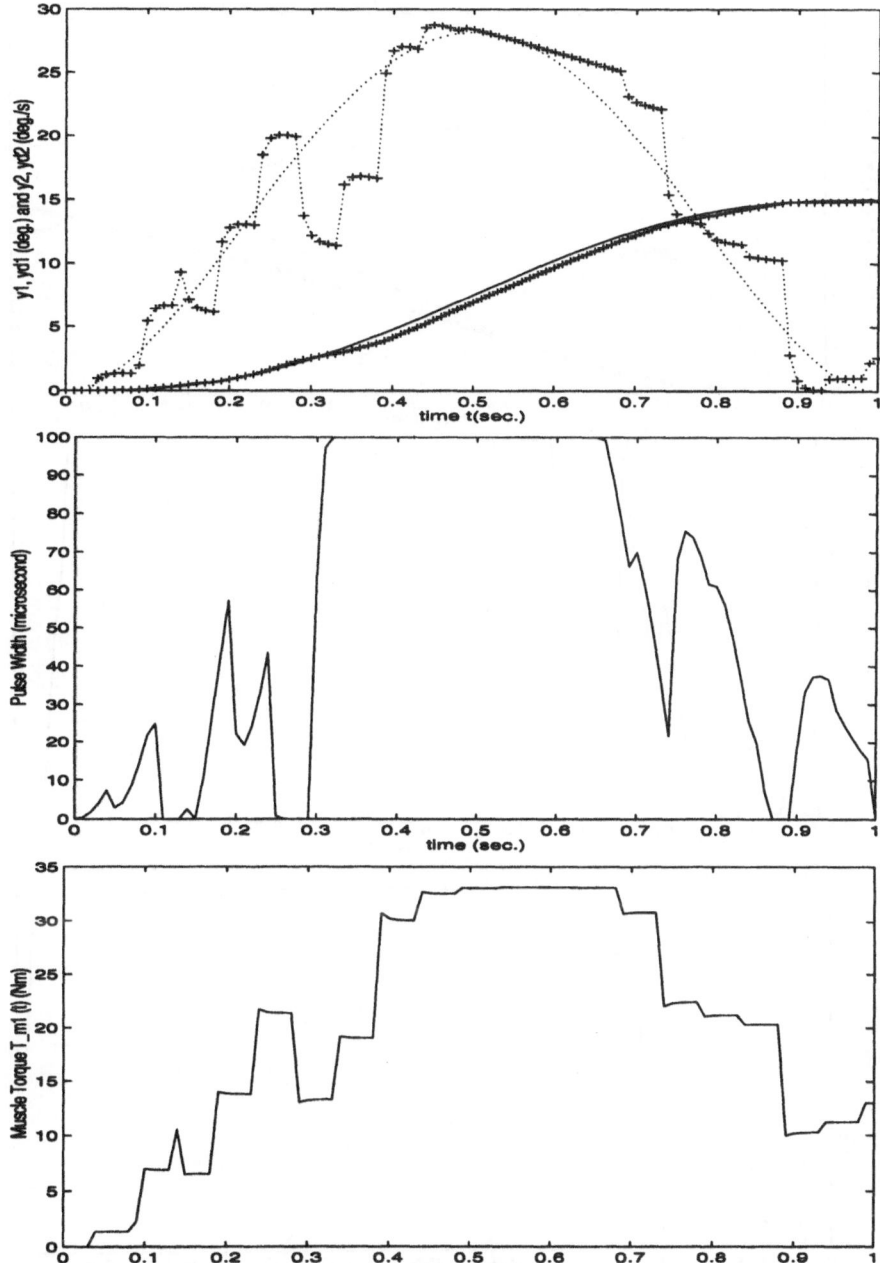

Fig. 11.16. Test: Responses of conventional PD controller, flexor only, $\theta_f = 15°$

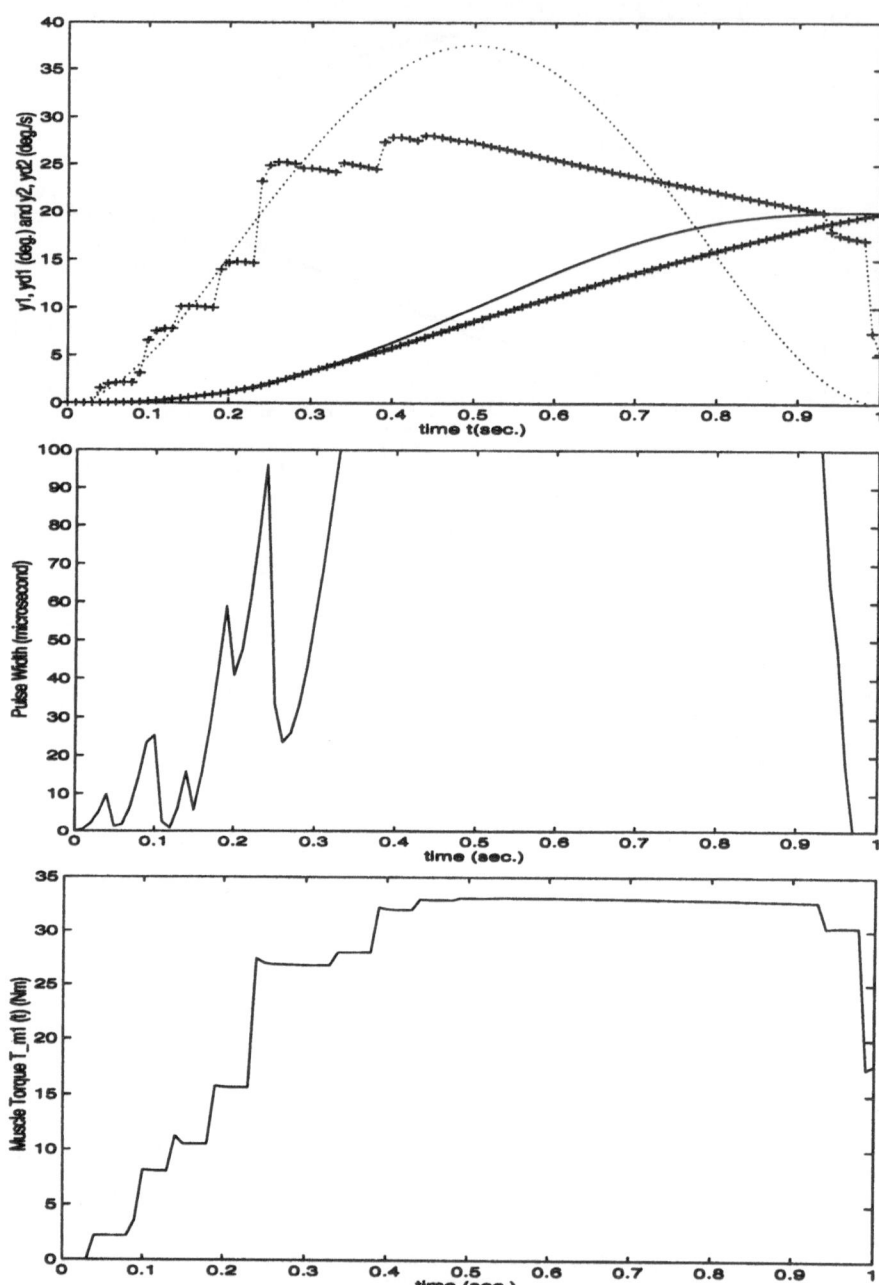

Fig. 11.17. Test: Responses of conventional PD controller, flexor only, $\theta_f = 20°$

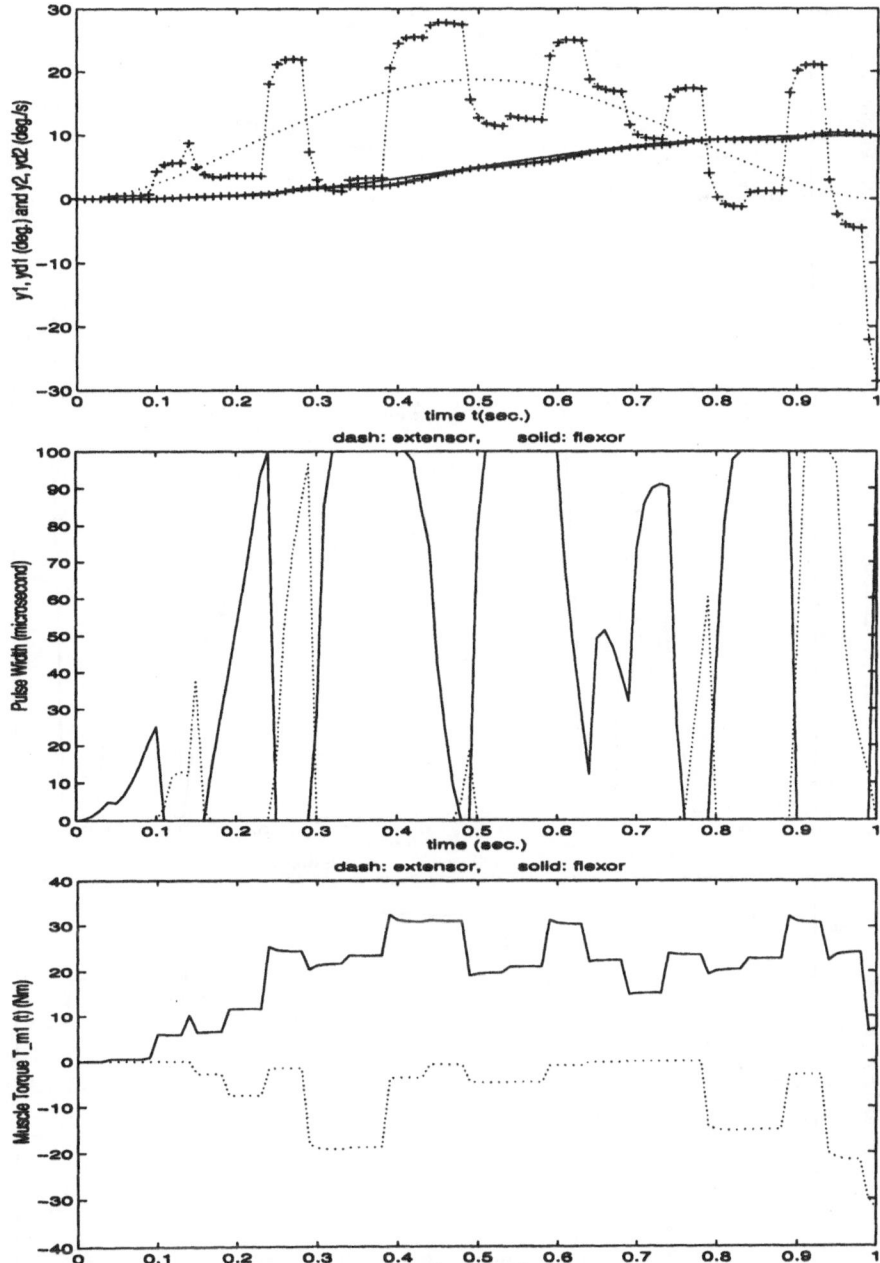

Fig. 11.18. Test: Responses of conventional PD controller, flexor and extensor, $\theta_f = 10°$

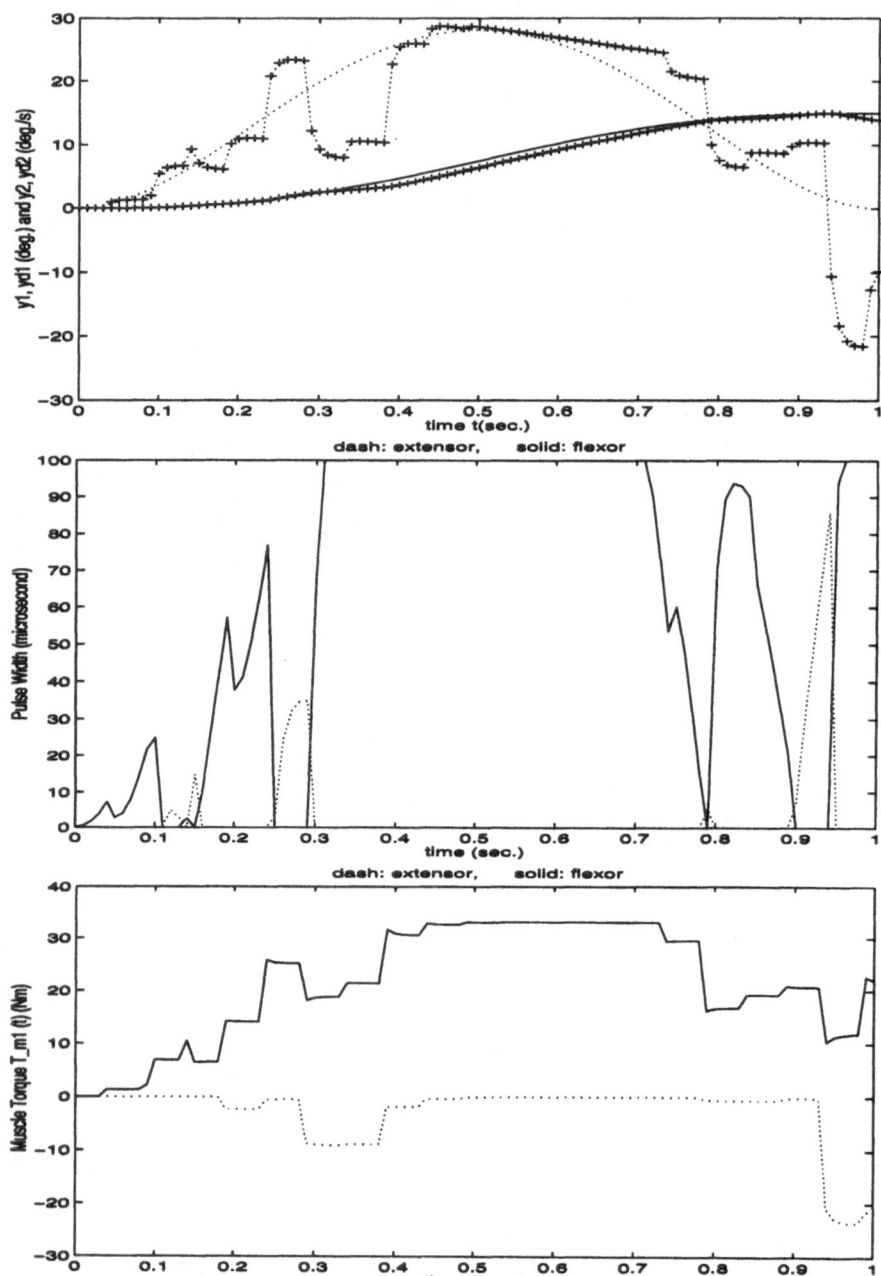

Fig. 11.19. Test: Responses of conventional PD controller, flexor and extensor, $\theta_f = 15°$

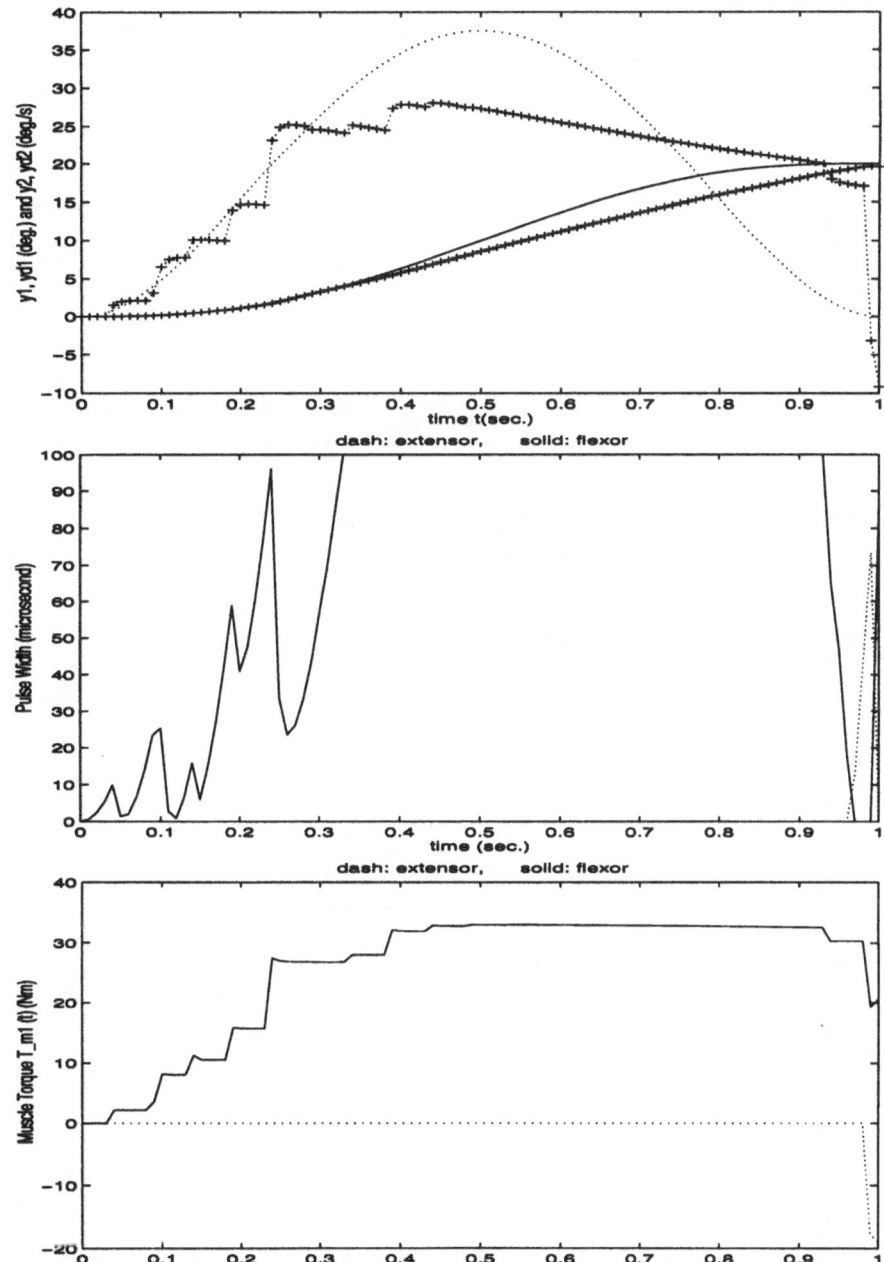

Fig. 11.20. Test: Responses of conventional PD controller, flexor and extensor, $\theta_f = 20°$

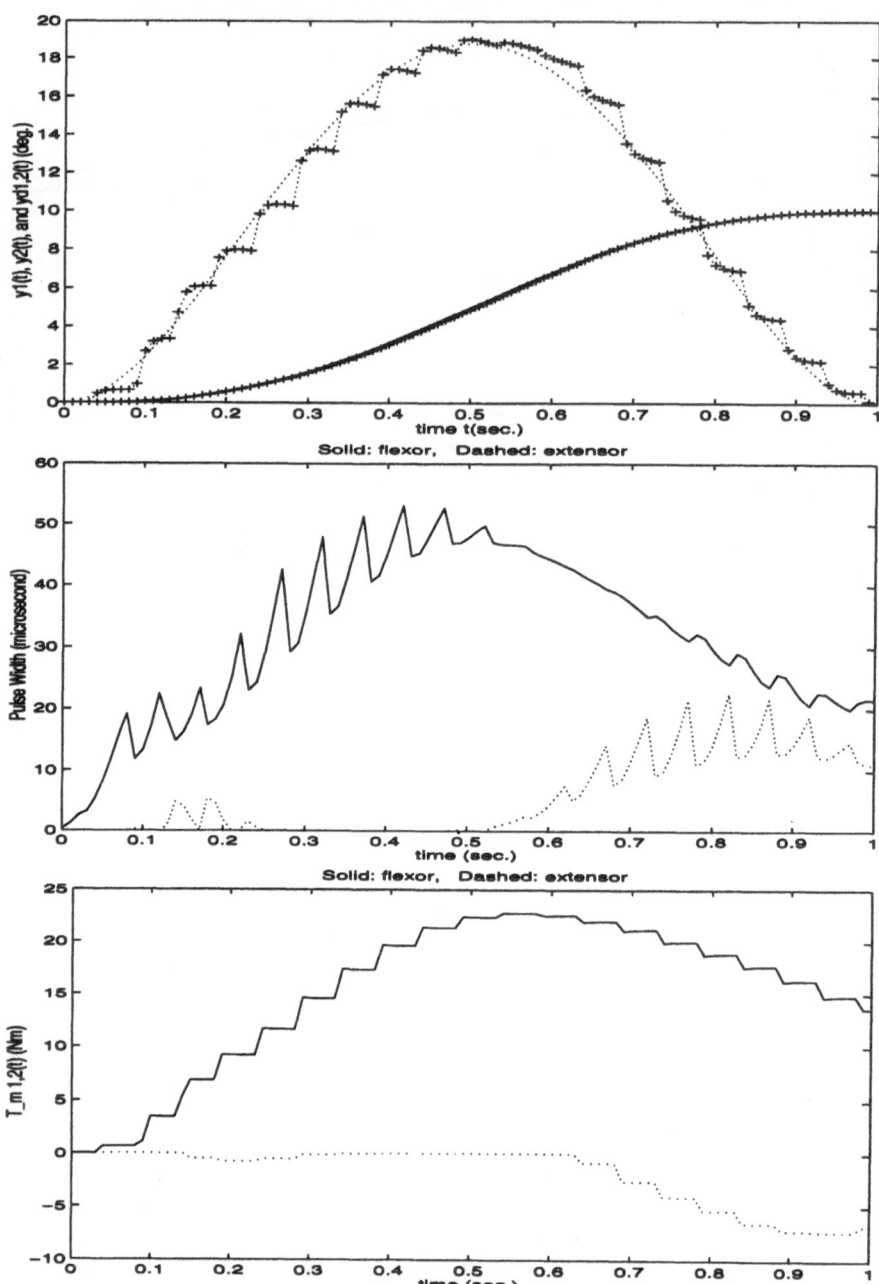

Fig. 11.21. Responses of high-order ILC at 12-th iteration (M=2)

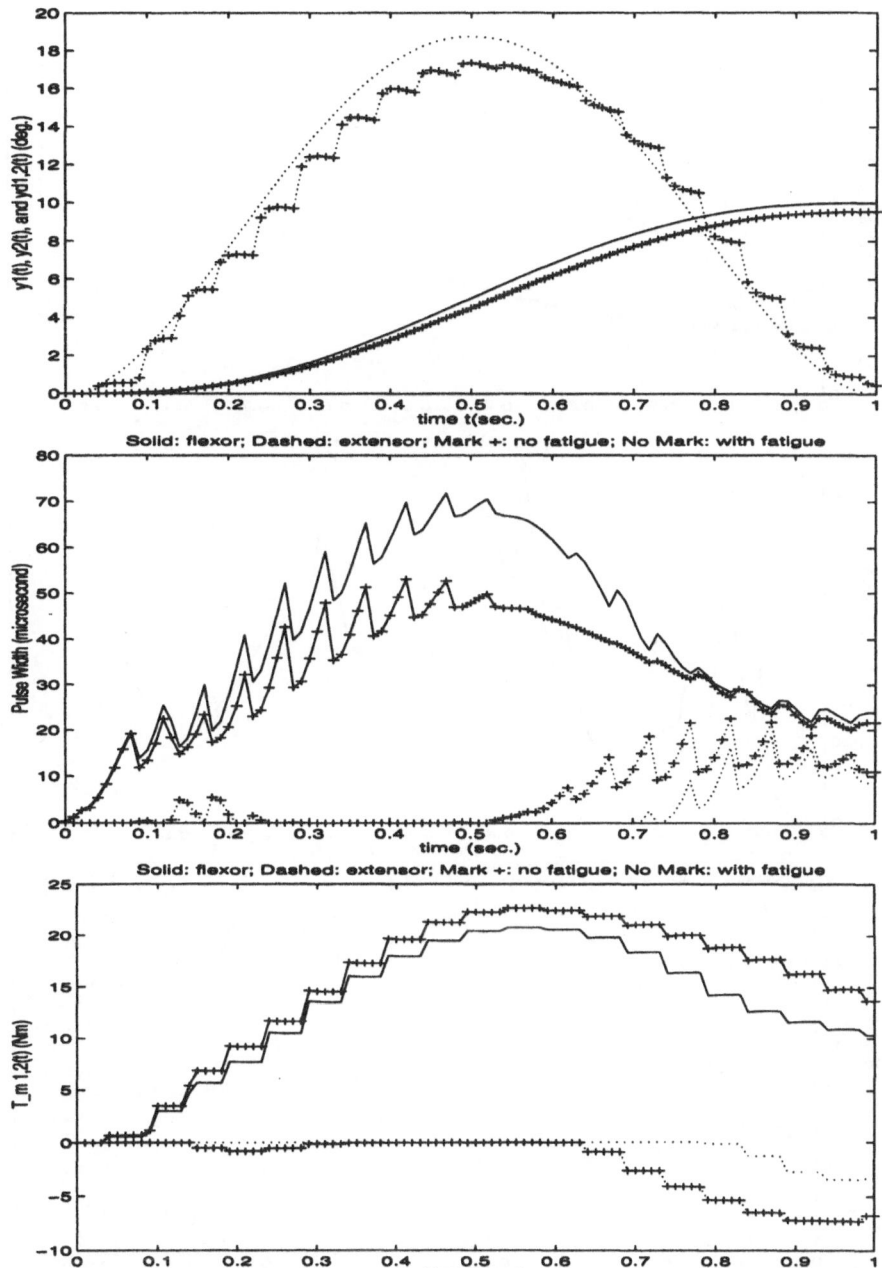

Fig. 11.22. Responses of high-order ILC at 12-th iteration with muscle fatigue effect (M=2)

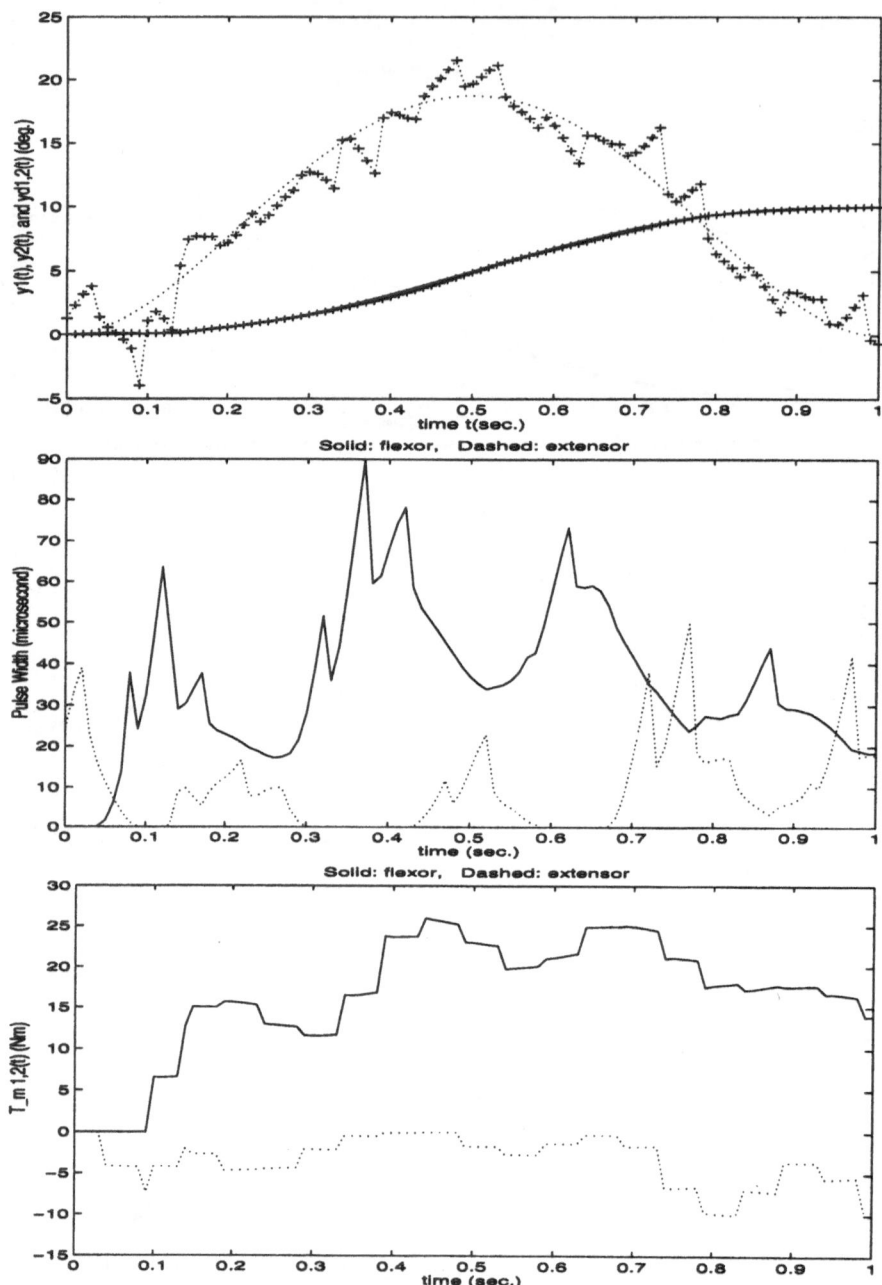

Fig. 11.23. Responses of high-order ILC at 12-th iteration with exogenous torque disturbance (M=2)

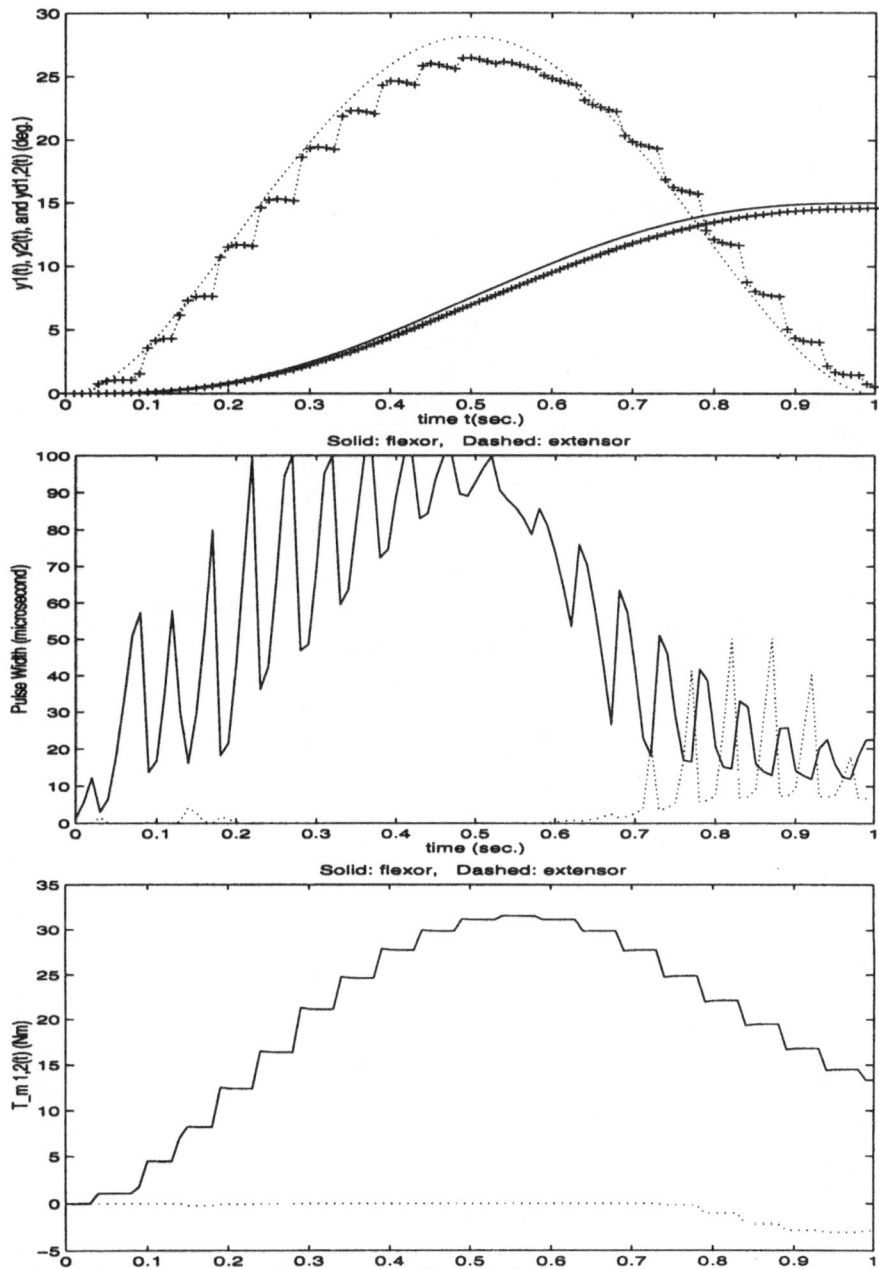

Fig. 11.24. Responses of high-order ILC at the 51st iteration with varying desired trajectories (M=2)

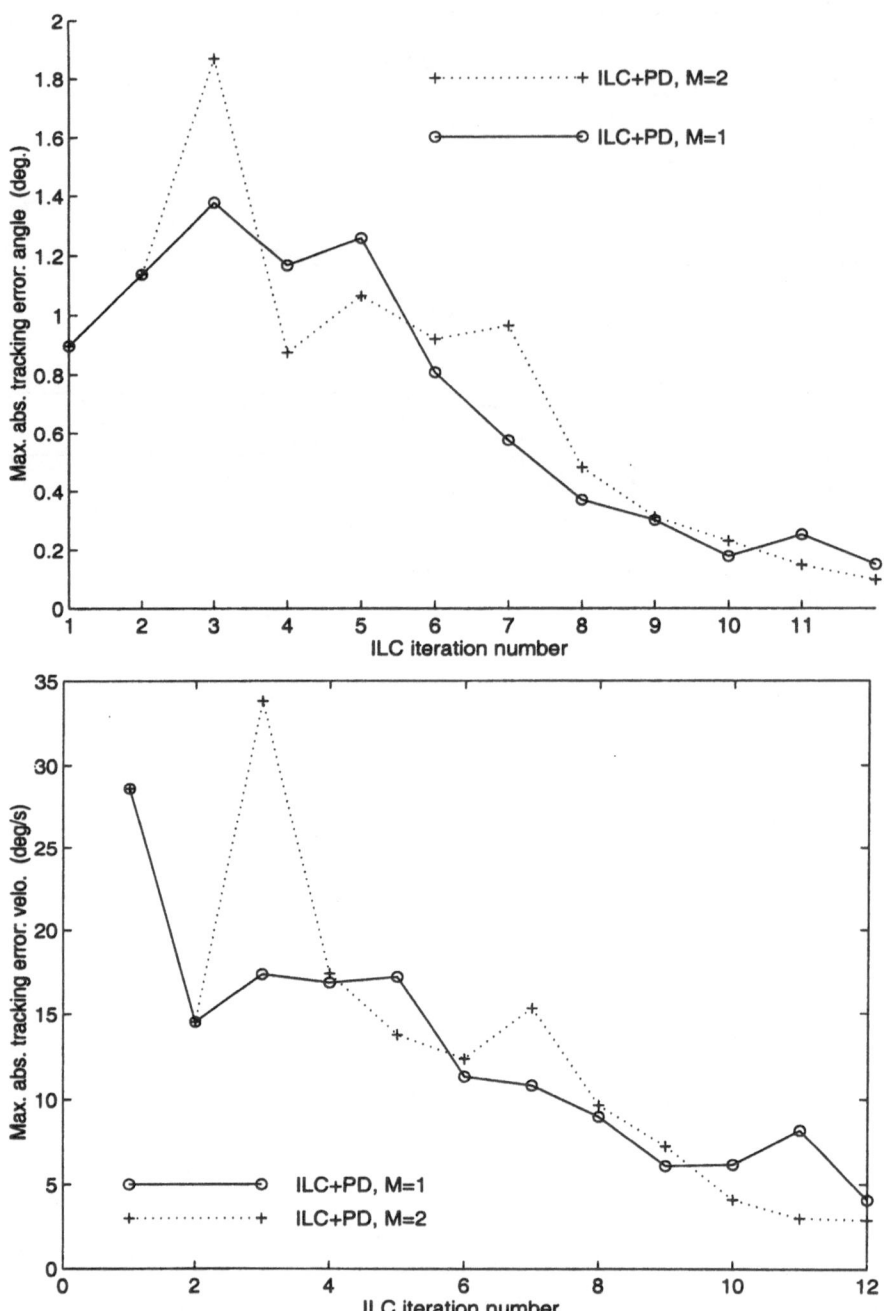

Fig. 11.25. ILC history of maximal absolute tracking errors comparison: joint angle and joint angular velocity

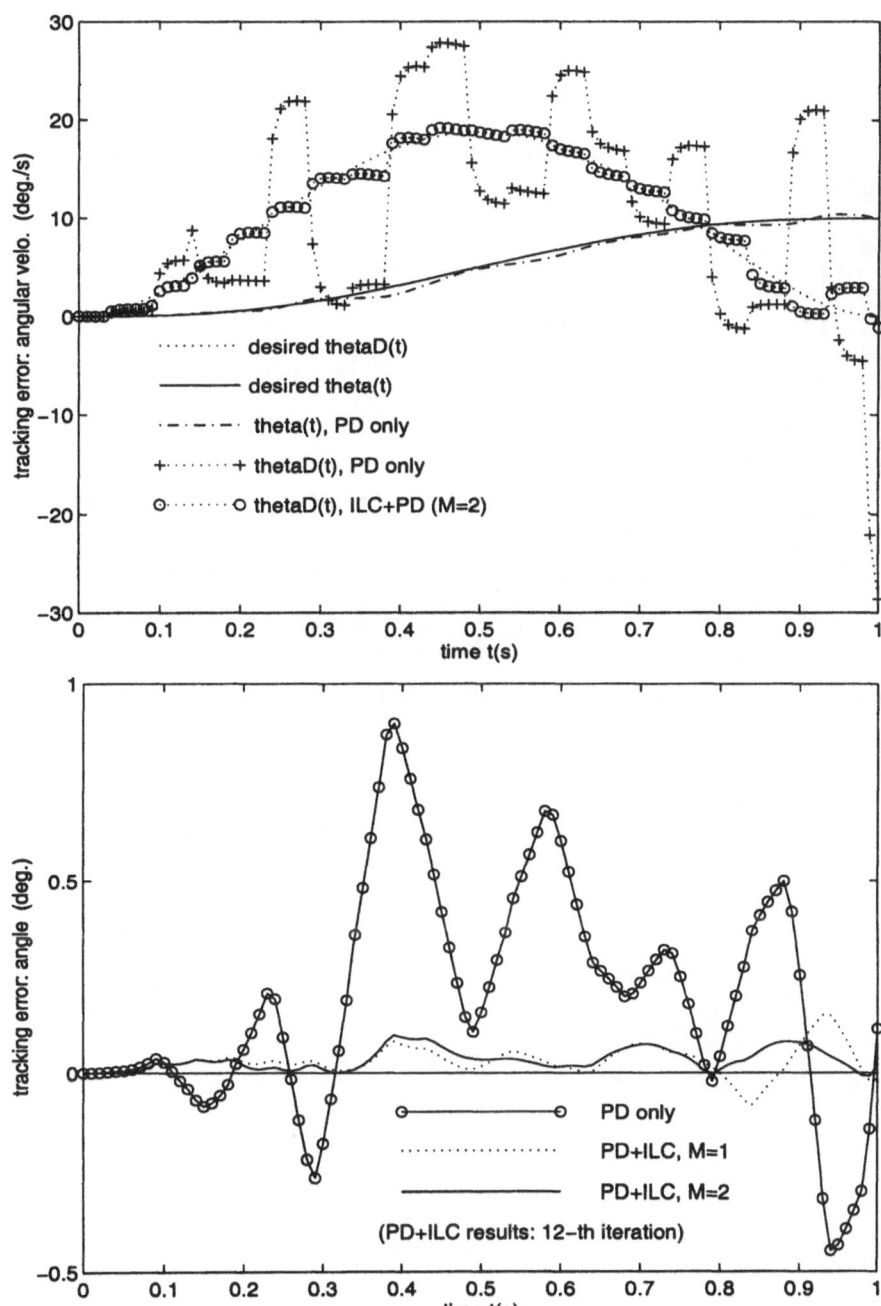

Fig. 11.26. Tracking error comparison: angular velocity and joint angle

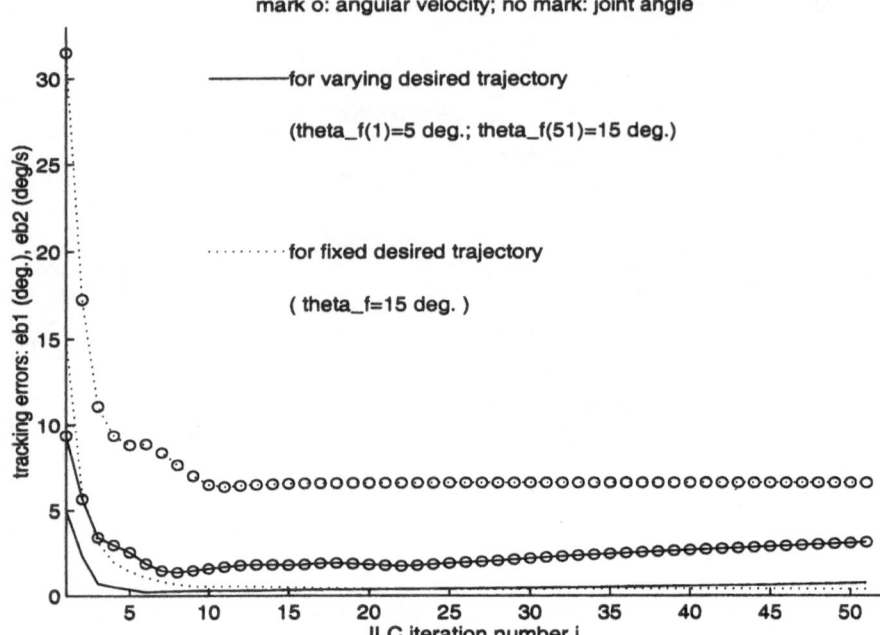

Fig. 11.27. ILC convergence comparison, varying desired trajectories

12. Conclusions and Future Research

12.1 Conclusions

SEVERAL NEW ASPECTS of Iterative Learning Control (ILC) have been addressed and investigated to show that improved control performance can be obtained by utilizing the system repetitions.

- A high-order ILC scheme is proposed and explained in the iteration number direction. By considering the dynamics along the ILC iteration number direction, the high-order scheme offers potentials to improve the ILC convergence property compared to the conventional ones which are merely in a pure integral controller form.
- In continuous-time systems, the current iteration tracking error information usage in Chapter 3 is shown to be helpful in tuning the tracking error bound and the ILC convergence rate.
- The ILC for uncertain discrete-time nonlinear systems has been studied in Chapters 4 and 5 where the effect of actuator saturation is considered. The feedback-assisted ILC and the ILC utilizing the current iteration tracking error are investigated separately in these two chapters and the differences are discussed explicitly.
- The ILC scheme with an iterative initial state learning method is shown to be effective to remove a commonly used re-initialization assumption in the conventional ILC methods. The unknown desired states can be identified through the ILC process. Extension to the case of a high-order ILC updating law is also discussed in Section 6.5.
- In Chapter 7, a terminal iterative learning control method is proposed for a point-to-point control problem where only the terminal output tracking error instead of the whole output trajectory tracking error is available. A high-order Terminal Iterative Learning Control method is proposed and applied to a typical terminal thickness control problem in RTP chemical vapor deposition (CVD) of wafer fab industry. A convergence condition is established for a class of uncertain discrete-time time-varying linear systems. Simulation results for an RTPCVD thickness control problem are presented to demonstrate the effectiveness of the proposed iterative learning scheme.

- The ILC *design* issues has been considered and a new design framework using noncausal filtering technique has been proposed in Chapter 8 for discrete-time case. Chapter 9 presents a parallel result to Chapter 8 for continuous-time situation. The convergence analysis shows that it is alway possible to ensure the ILC convergence with a proper choose of two parameters: the learning gain and the filtering length. Therefore, the *design* task of ILC is reduced to tuning these two parameters.
- Two new ILC applications, i.e., aerodynamic curve identification (Chapter 10) and neuromuscular stimulation control (Chapter 11), have been presented. These applications differ from the dominant applications of ILC to robotics. The results presented have demonstrated the effectiveness of the ILC methodology.

12.2 Future Research Directions

As discussed in Section 1.1.1, system repetitions provide additional information for the analysis of the repetitive system to be controlled. The basic rationale of Iterative Learning Control lies on the utilization of this repetition to update the future control actions in order to improve the system behavior. When a system performs a given task repeatedly, the ILC can be added in a 'plug-in' fashion to utilize this repetition to improve the control performance as the number of the repetitions increases. The existing controller of the original system need not to be changed. The robustness analysis of the overall system, which comprises ILC, existing feedforward controllers and feedback controllers, is an important direction of practical applications of the iterative learning control methodology. As the existing feedforward and feedback control strategies are abundant, the future research resource in the field of iterative learning control is clearly rich.

The recommended future research possibilities are listed as follows:

- *Learning Adaptive Control and Adaptive Learning Control.* As adaptive control has been received extensive research and found successful applications, the ILC scheme combined with the adaptive control will be attractive. Some attempts have been made in [169, 130, 185, 133, 122]. However, the adaptation is only in the time axis which should be called 'Learning Adaptive Control'. The non-identifier-based adaptive control methods [115, 116] are worthy noticing for the development of the Learning Adaptive Controls. For the adaptation in the ILC iteration number direction, a recursive identification algorithm is proposed in [87, 88] for linear time invariant system where improved results were observed. Some 'soft-computation-based' methods [259, 92] also addressed the 'adaptive learning' problem. But those are different from the ILC concept. The actual way we should go is in fact in a framework of the 2-D system theory [30]. The adaptation takes place in both the time axis and the ILC iteration number direction as presented in [5].

- *Learning Robust Control.* There are a lot of interests in the robust control of uncertain nonlinear systems, especially for the robot [2]. For recent advances, see [196, 149] and the references therein. All the controllers in this field are based on the Lyapunov function method. If the iterative learning controller is plugged to the existing robust controller, the analysis of the 'Learning Robust Control' is also quite interesting. The Lyapunov technique can still be used. However, it should be in the 2-D sense. A recently defined notion for this purpose can be found in [135].

- *Nonlinear ILC updating law.* In Chapter 10, a simple nonlinear ILC updating law – bi-linear scheme, is applied which is demonstrated to give an improved performance. However, there is no formal proposal for the effective ILC updating law in a nonlinear form. Actually, this can be regarded as the nonlinear controller design in the iteration number direction which can borrow some of the results from the existing discrete nonlinear feedback control schemes [37, 112, 113, 230, 137, 231, 224, 95].

- *Application Issues.* The ILC is a method of 'learning from practice'. This application-oriented methodology draws increasing attentions from the control community. However, the following practical issues are to be taken into consideration and are even challenging.

  - *P-type ILC.* When the system to be controlled has higher relative degree, say n, conventional D-type ILC theory requires the information of the n-th derivative of the tracking error. For example, in the one-link manipulator case, the second order derivative of the joint angle tracking error is required. Developing ILC algorithm using lower degree of tracking error derivative is economic and theoretically challenging. Remarkable attempts have been made in [218, 159, 185, 65]. The P-type ILC for the SPR (Strict Positive Realness) systems has been fully investigated in a recent work [131]. However, further studies are still in need for nonolinear systems.

  - *Control Saturation.* As explained in Chapter 2, the conventional ILC updating law is in an integral controller form when considering the dynamics along the ILC iteration number direction. An integral controller may windup under control input constraints which is the case in control practice [21]. This was observed in ILC applications and an anti-windup scheme is applied in [213]. However, no rigorous analysis considered in [213]. In the discrete-time case, a rigorous analysis of ILC convergence in the presence of the actuator saturation is given in Chapter 5. The simulation results for the application of ILC method to an FNS system with a maximal stimulation PW limitor demonstrate that ILC works well under hard saturation limit as shown in Chapter 11. It is conjectured that by limiting the control input at each ILC iteration on purpose might be helpful to the ILC performance enhancement. Moreover, by utilizing the existing schemes for the anti-windup design [91, 98, 97, 35, 29], more practical ILC schemes can be developed and analized.

- *Transient Shaping and Monitoring*. It has been observed that a transient period exists at the early stage of iterative learning control and when a disturbance runs in or out. In some applications, such transient must be restricted. To shape the desirable ILC transient, it is necessary to include a monitoring level in the iterative learning control. The monitoring level fuses the information in the previous iterations and acts mainly through the bounding of the control inputs. An idea in adaptive control to restrict the parameters in the parameter updating process is shown to be helpful which was explored in [245]. The author believes that the adaptively adjusted bound (not the maximum allowable bound) on the actual control input in ILC will surely be beneficial to the Transient Shaping and Monitoring.

- *ILC for Similar Tasks*. The ILC requires to learn to perfectly track a new given desired trajectory from beginning. If we already have several learned desired inputs for the desired trajetories, how to utilize them is an interesting problem. This was discussed in [126, 124, 125] using simply the idea of static interpolation. Recently, this is also investigated in [248, 191]. Further analytic result is in demand.

- *Systematic Design Methods*. Design issue is always problem specific. Currently, in ILC research, the learning gain is designed based on the ILC convergence condition which may not lead to a good design in terms of known knowledge assumed. Therefore, systematic design method is in great desire. Some local design methods are proven to be effective as reported in [192].

It should be pointed out at this moment that when incoporating the soft-computation methods like Neural Networks, Fuzzy Logic, Genetic Algorithm, Evolution Programming, CMAC, Associate Memory etc. into the Iterative Learning Control, a lot of new research horizons will appear. However, the currently explored good convergence properties of ILC which are fully supported by mathematics, will be attenuated. Even though we did not slip into the field of soft-computation, there are still so many aspects left unaddressed for such an obvious control rationale: *repetition improves skill, for either man and machine.*

References

1. J. J. Abbas and H. J. Chizeck. Neural network control of functional neuro-muscular stimulation systems: computer simulation studies. *IEEE Trans. on Biomedical Engineering*, 42(11):1117–1127, 1995.
2. C. Abdallah, D. Dawson, P. Dorato, and M. Jamshidi. Survey of robust control for rigid robots. *IEEE Control Systems*, pages 24–30, Feb. 1991.
3. H. S. Ahn, C. H. Choi, and K. B. Kim. Iterative learning control for a class of nonlinear systems. *Automatica*, 29(6):1575–1578, 1993.
4. H. S. Ahn, S.-H. Lee, and D.-H. Kim. Frequency-domain design of iterative learning controllers for feedback systems. In *IEEE International Symposium on Industrial Electronics*, volume 1, pages 352–357, Athens, Greece, July 1995.
5. G. M. Aly, H. Mahgoub, E. Zakzouk, and M. El-Assar. Design of 2-D model reference adaptive systems based on the second method of Lyapunov. *Control Theory and Technology*, 10(2):281–195, 1994.
6. N. Amann and D. H. Owens. Non-minimum phase plants in iterative learning control. In *Second International Conference on Intelligent Systems Engineering*, pages 107–112, Hamburg-Harburg, Germany, September 1994.
7. N. Amann and D. H. Owens. Iterative learning control for discrete time systems using optimal feedback and feedforward actions. In *Proceedings of the 34th Conference on Decision and Control*, New Orleans, LA, July 1995.
8. N. Amann, D. H. Owens, and E. Rogers. 2D systems theory applied to learning control systems. In *Proc. of the 33rd Conf. on Decision and Control*, pages 985–986, Lake Bruena, FL, USA, Dec. 1994.
9. N. Amann, D. H. Owens, and E. Rogers. Norm-optimal predictive iterative learning control. In *Proc. of European Control Conference*, pages 2880–2885, Rome, Italy, Sept. 1995.
10. N. Amann, D. H. Owens, and E. Rogers. Iterative learning control using optimal feedback and feedforward actions. *International Journal of Control*, 65(2):277–293, September 1996.
11. N. Amann, D. H. Owens, and E. Rogers. Robustness of norm-optimal iterative learning control. In *Proceedings of International Conference on Control*, volume 2, pages 1119–1124, Exeter, UK, September 1996.
12. N. Amann, D. H. Owens, E. Rogers, and A. Wahl. An H_∞ approach to linear iterative learning control design. *International Journal of Adaptive Control and Signal Processing*, 10(6):767–781, November-December 1996.
13. N. Amann, H. Owens, and E. Rogers. Iterative learning control for discrete-time systems with exponential rate of convergence. *IEE Proceedings-Control Theory and Applications*, 143(2):217–224, March 1996.
14. L. C. Anderson and J. H. Vincent. Application of system identification to aircraft flight test data. In *Proceedings of the 24th IEEE Conference on Decision and Control*, pages 1929–1931, Fort Lauderdale, FL, USA, Dec. 1985.

15. S. Arimoto. Mathematical theory of learning with applications to robot control. In K.S. Narendra, editor, *Adaptive and Learning Systems: Theory and Applications*, pages 379–388. Yale University, Yale University, New Haven, Connecticut, USA, Nov. 1985.

16. S. Arimoto. Robustness of learning control for robot manipulators. In *Proc. of the 1990 IEEE Int. Conf. on Robotics and Automation*, pages 1528–1533, Cincinnati, Ohio, USA, 1990.

17. S. Arimoto, S. Kawamura, and F. Miyazaki. Bettering operation of robots by learning. *J. of Robotic Systems*, 1(2):123–140, 1984.

18. S. Arimoto, S. Kawamura, and F. Miyazaki. Convergence, stability and robustness of learning control schemes for robot manipulators. In M. J. Jamishidi, L. Y. Luh, and M. Shahinpoor, editors, *Recent Trends in Robotics: Modelling, Control, and Education*, pages 307–316. Elsevier, NY, USA, 1986.

19. S. Arimoto and T. Naniwa. Learning control for motions under geometric endpoint constraint. In *Proc. of American Control Conference*, pages 2634–2638, 1992.

20. S. Arimoto, T. Naniwa, and H. Suzuki. Robustness of P-type learning control with a forgetting factor for robotic motions. In *Proc. of the 29-th IEEE Conf. on Decision and Control*, pages 2640–2645, Honolulu, Hawaii, USA, Dec. 1990.

21. K. J. Åström and L. Rundqwist. Integrator windup and how to avoid it. In *Proc. of American Control Conf.*, pages 1693–1698, Pittsbergh, PA, USA, June 1989.

22. C. G. Atkeson and J. McIntyre. Robot trajectory learning through practice. In *Proc. of IEEE ROBOT'86*, pages 1737–1742, San Francisco, CA, USA, 1986.

23. N. Bartelson. A method for determination of aerodynamic drag by doppler data. In *Proc. of the First Int. Ballistics Symp.*, Orlando, USA, 1975.

24. S. P. Bhat and D. K. Miu. Solutions to point-to-point control problem using Laplace transform technique. *ASME: J. of Dynamic Systems, Measurement, and Control*, 113:425–431, 1991.

25. S. P. Bhat and D. K. Miu. Point-to-point positioning of flexible structures using a time domain LQ smoothness constraint. *ASME: J. of Dynamic Systems, Measurement, and Control*, 114:416–421, 1992.

26. W. L. Bialkowski. Dreams versus reality: A view from both sides of the gap. *Pulp and Paper Canada*, 94(11), 1993.

27. Z. Bien and K. M. Huh. High-order iterative learning control algorithm. *IEE Proceedings, Part-D, Control Theory and Applications*, 136(3):105–112, 1989.

28. Z. Bien and J. - X. Xu, editors. *Iterative Learning Control - Analysis, Design, Integration and Applications*. Kluwer Academic Publishers, 1998.

29. C. Bohn and D. P. Atherton. An analysis package comparing PID anti-windup strategies. *IEEE Control Systems*, 15(2):34–40, 1995.

30. F. M. Boland and D. H. Owens. Linear multipass processes - a two-dimensional interpretation. *Proc. IEE*, 127(5):189–193, 1980.

31. P. Bondi, G. Casalino, and L. Gambardella. On the iterative learning control theory for robotic manipulators. *IEEE J. of Robotics and Automation*, 4:14–22, Feb. 1988.

32. Gary M. Bone. A novel iterative learning control formulation of generalized predictive control. *Automatica*, 31(10):1483–1487, 1995.

33. E. Burdet, L. Rey, and A. Codourey. A trivial method of learning control. In *Preprints of the 5th IFAC Symposium on Robot Control, volume 2*, Nante, France, 1997. IFAC.

34. G. Cai, Q. Zhou, and T. P. Leung. On learning control of constrained robots modeled by singular systems. *Control Theory and Applications*, 11(4):396–403, 1994.

35. P. J. Campo, M. Morari, and C. N. Nett. Multivariable anti-windup and bump-less transfer: a general theory. In *Proc. of American Control Conf.*, pages 1706–1711, Pittsbergh, PA, USA, June 1989.

36. G. Casalino and B. Bartolini. A learning procedure for the control of movements of robotic manipulators. In *Proc. of the IASTED Symp. on Robotics and Automation*, pages 108–111, Amsterdam, Netherland, 1984.

37. B. Castillo and S. D. Gennaro. Asymptotic output tracking for SISO nonlinear discrete-time systems. In *Proc. of the 30th Conf. on Decision and Control*, pages 1802–1806, Brighton, England, Dec. 1991.

38. G. T. Chapman and D. B. Kirk. A method for extracting aerodynamic coefficients from free flight data. *AIAA Journal*, 18(4), 1970.

39. C. C. Cheah and D. Wang. Learning control for a class of nonlinear differential-algebraic systems with application to constrained robots. In *Proc. of American Control Conference*, pages 1737–1741, Baltimore, Maryland, USA, Jun 1994.

40. C. C. Cheah and D. Wang. Learning impedance control for robotic manipulators. In *Proc. of the 1995 IEEE Int. Conf. on Robotics and Automation*, 1995.

41. C. C. Cheah, D. Wang, and Y. C. Soh. Learning control of constrained robots. In *Proc. of the 1992 IEEE Int. Symp. on Intelligent Control*, pages 91–96, Aug. 1992.

42. C. C. Cheah, D. Wang, and Y. C. Soh. Convergence and robustness of a discrete-time learning control scheme for constrained manipulators. *J. of Robotic Systems*, 11(3):223–238, 1994.

43. Y. Chen. Researches on trajectory prediction models and software for spin-stabilized projectiles (I), (II), (III). *Projectile and Rocket Fascicule of Acta Armamentarii*, 90,1(1990(3):1-7; 1990(4):1-11; 1991(1):1-10 (in Chinese)), 1990,1.

44. Y. Chen and H. Dou. Researches on the optimal control solution of identifying fitting drag coefficient curve from radar measured velocity data. *Aerodynamic Experiment and Measurement Control*, 1993(2):81–89 (in Chinese), 1993.

45. Y. Chen and H. Dou. Robust curve identification by iterative learning. In *Proc. of the First Chinese World Congress on Intelligent Control and Intelligent Automation (CWC ICIA'93)*, pages 1973–1980, Beijing, China, Aug. 1993.

46. Y. Chen, Z. Gong, and C. Wen. Analysis of a high order iterative learning control algorithm for uncertain nonlinear systems with state delays. *Automatica*, 34(3):345–353, March 1998.

47. Y. Chen, D. Lu, H. Dou, and Y. Qing. Optimal dynamic fitting and identification of aerobomb's fitting drag coefficient curve. In *Proc. of the First IEEE Conf. on Control Applications*, pages 853–858, Dayton, Ohio, USA, Sept. 1992.

48. Y. Chen, M. Sun, and H. Dou. Dual-staged P-type iterative learning control schemes. In *Proc. of the First Asian Control Conference*, pages 239–242, Tokyo, Japan, July 1994.

49. Y. Chen, M. Sun, B. Huang, and H. Dou. Robust higher order repetitive learning control algorithm for tracking control of delayed repetitive systems. In *Proc. of the 31st IEEE Conf. on Decision and Control*, pages 2504–2510, Tucson, Arizona, USA, Dec. 1992.

50. Y. Chen, C. Wen, H. Dou, and M. Sun. Iterative learning identification of aerodynamic drag curve from tracking radar measurements. *Journal of Control Engineering Practice*, 5(11):1543–1554, Nov. 1997.

51. Y. Chen, C. Wen, Z. Gong, and M. Sun. An iterative learning controller with initial state learning. *IEEE Trans. on Automatic Control*, 44(2):371-376, Feb. 1999.

52. Y. Chen, C. Wen, Z. Gong, and M. Sun. Drag coefficient curve identification of projectiles from flight tests via optimal dynamic fitting. *Journal of Control Engineering Practice*, 5(5):627–636, May 1997.

53. Y. Chen, C. Wen, and M. Sun. A robust high-order P-type iterative learning controller using current iteration tracking error. *Int. J. of Control*, 68(2):331–342, Sept. 1997.

54. Y. Chen, C. Wen, J.-X. Xu, and M. Sun. High-order iterative learning identification of projectile's aerodynamic drag coefficient curve from radar measured velocity data. In *Proc. of the 35-th IEEE CDC*, pages 3070–1, Kobe, Japan, Dec. 1996.

55. Y. Chen, C. Wen, J.-X. Xu, and M. Sun. An initial state learning method for iterative learning control of uncertain time-varying systems. In *Proc. of the 35-th IEEE CDC*, pages 3996–4001, Kobe, Japan, Dec. 1996.

56. Y. Chen, C. Wen, J.-X. Xu, and M. Sun. High-order iterative learning identification of projectile's aerodynamic drag coefficient curve from radar measured velocity data. *IEEE Transaction on Control Systems Technology*, 6(4):563-570, Sept. 1998.

57. Y. Chen, J.-X. Xu, and T. H. Lee. Current iteration tracking error assisted iterative learning control of uncertain nonlinear discrete-time systems. In *Proc. of the 35th IEEE Conference on Decision and Control*, pages 3040–5, Kobe, Japan, Dec. 1996.

58. Y. Chen, J.-X. Xu, and T. H. Lee. Feedback-assisted high-order iterative learning control of uncertain nonlinear discrete-time systems. In *Proc. of the Int. Conf. on Control, Automation, Robotics and Vision (ICARCV)*, pages 1785–9, Singapore, Dec. 1996.

59. Y. Chen, J.-X. Xu, and T. H. Lee. An iterative learning controller using current iteration tracking error information and initial state learning. In *Proc. of the 35-th IEEE CDC*, pages 3064–9, Kobe, Japan, Dec. 1996.

60. Y. Chen, J.-X. Xu, and T. H. Lee. Current iteration tracking error assisted high-order iterative learning control of discrete-time uncertain nonlinear systems. In *Proc. of the 2nd Asian Control Conference (ASCC'97), Vol. I*, pages 573–6, Seoul, Korea, Jul. 1997.

61. Y. Chen, J.-X. Xu, T. H. Lee, and S. Yamamoto. An iterative learning control in rapid thermal processing. In *Proc. the IASTED Int. Conf. on Modeling, Simulation and Optimization (MSO'97)*, pages 189–92, Singapore, Aug. 1997.

62. Y. Chen, T. H. Lee, J.-X. Xu and S. Yamamoto, Noncausal filtering based design of iterative learning control. In K. L. Moore, editor, *Proc. of the First Int. Workshop on Iterative Learning Control*, pages 63–70, Hyatt Regency, Tampa, FL, USA, Dec. 1998.

63. Y. Chen, C. Wen, and M. Sun. Discrete-time iterative learning control of uncertain nonlinear feedback systems. In *Proc. of The Second Chinese World Congress on Intelligent Control and Intelligent Automation (CWC ICIA'97), Xi'an Jiaotong University Press.*, pages 1972–7, Xi'an, P.R. China, June 23-27 1997.

64. Y. Chen, C. Wen, and M. Sun. A high-order iterative learning controller with initial state learning. In *Proc. of The Second Chinese World Congress on Intelligent Control and Intelligent Automation (CWC ICIA'97), Xi'an Jiaotong University Press.*, pages 684–9, Xi'an, P.R. China, June 23-27 1997.

65. C.-J. Chien and J.-S. Liu. A P-type iterative learning controller for robust output tracking of nonlinear time-varying systems. In *Proc. of American Control Conference*, pages 2595–2599, Baltimore, Maryland, USA, Jun. 1994.

66. C.-H. Choi and T.-J. Jang. Iterative learning control for a general class of nonlinear feedback systems. In *Proceedings of the 1995 American Control Conference*, pages 2444–2448, Seattle, WA, June 1995.

67. J.-W. Choi, H.-G. Choi, K.-S. Lee, and W.-H. Lee. Control of ethanol concentration in a fed-batch cultivation of *acinetobacter calcoaceticus* RAG-1 using a

feedback-assisted iterative learning algorithm. *Journal of Biotechnology*, 49:29–43, August 1996.

68. P. I. Corke and B. Armstrong-Hélouvry. A meta-study of PUMA 560 dynamics: a critical appraisal of literature data. *Robotica*, 13:253–258, 1995.

69. P. E. Crago, N. Lan, P. H. Veltink, J. J. Abbas, and C. Kantor. New control strategies for neuroprosthetic systems. *Journal of Rehabilitation Research and Development*, 33(2):158–172, 1996.

70. J. J. Craig. Adaptive control of manipulators through repeated trials. In *Proc. of American Control Conference*, pages 1566–1573, San Diego, CA, USA, Jun. 1984.

71. D. Dawson, R. Genet, and F. L. Lewis. A hybrid adaptive/learning controller for a robot manipulator. In N. Sadegh, editor, *Adaptive and Learning Control*, pages 51–54. ASME, Dallas, TX, USA, Nov. 1990. The Winter Annual Meeting of the ASME.

72. D. de Roover. Synthesis of a robust iterative learning controller using an H_∞ approach. In *Proceedings of the 35th IEEE Conference on Decision and Control*, Kobe, Japan, December 1996.

73. T.-Y. Doh, K. B. Jin, and M. J. Chung. An LMI approach to iterative learning control for uncertain linear systems. In *ISIAC*, pages 006.1–006.6, Albuquerque, NM, USA, 1998. TSI Press.

74. H. Dou, Y. Chen, and M. Sun. Iterative learning identification of a nonlinear function in a nonlinear dynamic system. In *Proc. of the First IFAC Youth Automation Conference (IFAC YAC'95)*, pages 138–143, Beijing, China, Aug. 1995.

75. H. Dou, Z. Zhou, Y. Chen, J.-X. Xu, and J. Abbas. Iterative learning control strategy for functional neuromuscular stimulation. In *Proc. of the 1996 IEEE EMBS Int. Conf.*, Amsterdam, Oct. 1996.

76. H. Dou, Z. Zhou, Y. Chen, J.-X. Xu, and J. Abbas. Robust motion control of electrically stimulated human limb via discrete-time high-order iterative learning scheme. In *Presented at the 1996 Int. Conf. on Automation, Robotics and Computer Vision (ICARCV'96)*, pages 1087–91, Singapore, Dec. 1996.

77. H. Dou, Z. Zhou, Y. Chen, J.-X. Xu, and J. Abbas. Robust control of functional neuromuscular stimulation system by discrete-time iterative learning scheme. In *Proc. of the 2nd Asian Control Conference (ASCC'97), Vol. I*, pages 565–8, Seoul, Korea, Jul. 1997.

78. H. Dou, Z. Zhou, M. Sun, and Y. Chen. Robust high-order P-type iterative learning control for a class of uncertain nonlinear systems. In *Presented at the 1996 IEEE Int. Conf. on Systems, Man, and Cybernetics*, pages 923–928, Beijing, Oct. 1996.

79. J. B. Edwards and D. H. Owens. Stability problems in the control of of linear multipass processes. *Proc. IEE*, 121(11):1425–1431, 1974.

80. J. B. Edwards and D. H. Owens. *Analysis and Control of Multipass Processes*. Research Studies Press, Taunton, Chichester, 1982.

81. H. Elci, R. W. Longman, M. Phan, J. N. Juang, and R. Ugoletti. Discrete frequency based learning control for precision motion control. In *Proceedings of the 1994 IEEE International Conference on Systems, Man, and Cybernetics*, volume 3, pages 2767–2773, San Antonio, TX, 1994.

82. H. Elci, R. W. Longman, M. Phan, J.-N. Juang, and R. Ugoletti. Discrete frequency based learning control for precise motion control. In *Proc. of the IEEE Int. Conf. on System, Man, and Cybernetics*, pages 2767–2773, San Antonio, TX, USA, Oct. 1994.

83. H. Elci, H. N. Juang R. W. Longman, M. Phan, and R. Ugoletti. Automated learning control through model updating for precision motion control. *Adaptive*

Structures and Composite Materials: Analysis and Applications, ASME, AD-45/MD-54:299–314, 1994.

84. H. Elci. *The design and applications of learning control*. PhD thesis, Columbia University, 1995.

85. C. Fratter and R. F. Stengel. Identification of aerodynamic coefficients using flight testing data. *AIAA Paper No. 83-2099*, 1983.

86. A. Gautam. 2-D approximation and learning control of robot manipulators. Master's thesis, Florida Atlantic University, 1989.

87. Z. Geng. *Two dimensional model and analysis for a class of iterative learning control systems*. PhD thesis, George Washington University, 1990.

88. Z. Geng, R. Carroll, M. Jamshidi, and R. Kisner. A learning control scheme with gain estimator. In *Proc. of the 1991 IEEE Int. Symp. on Intelligent Control*, pages 365–370, Arlington, Virginia, USA, Aug. 1991.

89. Z. Geng, R. L. Carroll, and J. Xie. Two-dimensional model algorithm analysis for a class of iterative learning control systems. *International Journal of Control*, 52:833–862, 1990.

90. Z. Geng, J. D. Lee, R. L. Carroll, and L. H. Haynes. Learning control system design based on 2-D theory - an application to parallel link manipulator. In *Proc. of the 1990 IEEE Int. Conf. on Robotics and Automation*, pages 1510–1515, 1990.

91. A. H. Glattfelder and W. Schaufelberger. Stability analysis of single loop control systems with saturation and antireset-windup circuits. *IEEE Trans. of Automat. Contr.*, 28(12):1074–1081, 1986.

92. D. M. Gorinevsky. Adaptive learning control using affine radial basis function network approximation. In *Proc. of the 1993 IEEE Int. Symp. on Intelligent Control*, pages 505–510, Chicago, IL, USA, Aug. 1993.

93. N. K. Gupta and K. W. Illif. Identification of aerodynamic indicial functions using flight data. *AIAA Paper 82-1375*, 1982.

94. R. S. Gyurcsik, T. J. Riley, and F. Y. Sorrell. A model for rapid thermal processing: achieving uniformity through lamp control. *IEEE Trans. on Semiconductor Manufacturing*, 6(4):9–13, 1991.

95. E. Gyurkovics and T. Takacs. Exponential stabilization of discrete-time uncertain systems under control constraints. In *Proc. of the European Control Conf.*, pages 3636–3641, Rome, Italy, Sept. 1995.

96. R. D. Hanson and T.-C. Tsao. Compensation for cutting force induced bore cylindricity dynamic errors - a hybrid repetitive servo/iterative learning process feedback control approach. In *Proceedings of the Japan/USA Symposium on Flexible Automation*, volume 2, pages 1019–1026, Boston, MA, July 1996.

97. R. Hanus and M. Kinnaert. Control of constrained multivariable systems using the conditioning techniques. In *Proc. of American Control Conf.*, pages 1712–1718, Pittsbergh, PA, USA, June 1989.

98. R. Hanus, M. Kinnaert, and J.-L. Henrotte. Conditioning technique, a general anti-windup and bumpless transfer method. *Automatica*, 23(6):729–739, 1987.

99. E. G. Harokopos. Learning and optimal control of industrial robots in repetitive motions. In *Technical Paper No. MS86-387 of Robots 10 Conf., Society of Manufacturing Engineers*, pages 429–446, Chicago, IL, USA, Apr. 1986.

100. H. Hashimoto and J.-X. Xu. Learning control systems with feedback. In *Proceedings of the IEEE Asian Electronics Conference*, Hong Kong, September 1987.

101. J. Hauser. Learning control for a class of nonlinear systems. In *Proc. of the 26-th IEEE Conf. on Decision and Control*, pages 859–860, Los Angles, CA, USA, Dec. 1987.

102. G. Heinzinger, D. Fenwick, B. Paden, and F. Miyazaki. Robust learning control. In *Proc. of the 28-th IEEE Conf. on Decision and Control*, pages 436–440, Tempa, FL, USA, Dec. 1989.

103. G. Heinzinger, D. Fenwick, B. Paden, and F. Miyazaki. Stability of learning control with disturbances and uncertain initial conditions. *IEEE Trans. of Automatic Control*, 37(1):110–114, 1992.

104. L. Hideg and R. Judd. Frequency domain analysis of learning systems. In *Proceedings of the 27th Conference on Decision and Control*, pages 586–591, Austin, Texas, December 1988.

105. L. Hideg and R. Judd. Coefficient test for discrete time learning system stability. In *Proc. of the 1991 IEEE Int. Symp. Intelligent Control*, pages 383–387, Arlington, Virginia, USA, Aug. 1991.

106. L. M. Hideg. Time delays in iterative learning control schemes. In *Proceedings of the 1995 IEEE International Symposium on Intelligent Control*, pages 5–20, Monterey, CA, August 1995.

107. G. Hillerström and J. Sternby. Rejection of perodic disturbances with unknown period - a frequency domain approach. In *Proc. of American Control Conference*, pages 1626–1631, San Francisco, CA, USA, Jun. 1993.

108. G. Hillerström. Adaptive suppression of vibrations-a repetitive control approach. *IEEE Transactions on Control Systems Technology*, 4(1), January 1996.

109. R. Horowitz. Learning control of robot manipulators. *ASME Transactions: J. of Dynamic Systems, Measurement, and Control*, 115:402–411, 1993.

110. R. Horowitz, W. Messner, and M. Boals. Exponential convergence of a learning controller for robot manipulator. *IEEE Trans. of Automatic Control*, 36(7):890–894, 1991.

111. J.-S. Hu and M. Tomizuka. Adaptive asymptotic tracking of repetitive signals - a frequency domain approach. *IEEE Transactions on Automatic Control*, 38(19), October 1993.

112. J. Huang and C.-F. Lin. On the discrete-time nonlinear servomechanism problems. In *Proc. of American Control Conf.*, pages 1802–1806, San Francisco, California, USA, June 1993.

113. J. Huang and C.-F. Lin. A stability property and its application to discrete-time nonlinear system control. *IEEE Trans. of Automat. Contr.*, 39(11):2307–2311, 1995.

114. D.-H. Hwang, Z. Bien, and S.-R. Oh. Iterative learning control method for discrete-time dynamic systems. *IEE Proceedings, Part-D, Control Theory and Applications*, 138(2):139–144, 1991.

115. A. Ilchmann. Non-identifier-based adaptive control of dynamic systems: a survey. *IMA J. of Math. Control and Infomation*, 8:321–366, 1992.

116. A. Ilchmann. *Non-identifier-based high-gain adaptive control.* Lecture Notes in Control and Information Sciences: Vol. 189. London; New York : Springer-Verlag, 1993.

117. P. A. Ioannou and J. Sun. *Robust Adaptive Control.* Prentice-Hall International, Inc., 1996.

118. T. Ishihara, K. Abe, and T. Takeda. A discrete-time design of robust iterative learning controller. *IEEE Trans. on Systems, Man and Cybernetics*, 22(1):74–84, 1992.

119. T.-J. Jang, H.-S. Ahn, and C.-H. Choi. Iterative learning control for discrete-time nonlinear systems. *International Journal of Systems Science*, 25(7):1179–1189, 1994.

120. T.-J. Jang, C.-H. Choi, and H.-S. Ahn. Iterative learning control in feedback systems. *Automatica*, 31(2):243–245, 1995.

121. Q. Jiang and Q. Chen. Dynamic model for real-time estimation of aerodynamic characteristics. In *Collection of Technical Papers - AIAA Atmospheric Flight Mechanics Conference*, pages 331–339, Williamsburg, VA, USA, Aug. 1986.

122. Y. A. Jiang, D. J. Clements, T. Hesketh, and J. S. Park. Adaptive learning control of robot manipulators in task space. In *Proc. of American Control Conference*, pages 207–211, Baltimore, Maryland, USA, Jun. 1994.

123. R. P. Judd, R. P. Van Til, and L. Hideg. Equivalent Lyapunov and frequency domain stability conditions for iterative learning control systems. In *Proceedings of 8th IEEE International Symposium on Intelligent Control*, pages 487–492, 1993.

124. S. Kawamura and N. Fukao. Interpolations for input torque patterns obtained through learning control. In *Proc. of the 3rd Int. Conf. on Automation, Robotics and Computer Vision (ICARCV'94)*, pages 2194–2198, Singapore, Nov. 9-11 1994.

125. S. Kawamura and N. Fukao. A time-scale interpolation for input torque patterns obtained through learning control on constrained robot motions. In *Proc. of the 1995 IEEE Int. Conf. on Robotics and Automation*, pages 2156–2161, 1995.

126. S. Kawamura, F. Miyazaki, and S. Arimoto. Realization of a desired motion based on learning method and use of knowledge obtained from learning. In *SICE'87*, pages 953–956 (in Japanese), Hiroshima, July 15-17 1987.

127. S. Kawamura and N. Fukao. Use of feedforward input patterns obtained through learning control. In *Proceedings of the 2nd Asian Control Conference*, Seoul, Korea, July 1997.

128. C. Kempf, W. Messner, M. Tomizuka, and R. Horowitz. A comparison of four discrete-time repetitive control algorithms. In *Proc. of American Control Conference*, pages 2700–2704, 1992.

129. B. K. Kim, W. K. Chung, and Y. Youm. Robust learning control for robot manipulators based on disturbance observer. In *IECON Proceedings (Industrial Electronics Conference)*, volume 2, pages 1276–1282, Taipei, Taiwan, 1996.

130. T.-Y. Kuc and J. S. Lee. An adaptive learning control of uncertain robotic systems. In *Proc. of the 30th Conf. on Decision and Control*, pages 1206–1211, Brighton, England, Dec. 1991.

131. T.-Y. Kuc and J.-S. Lee. Learning strictly positive real linear systems with uncertain parameters and unknown input disturbances. *Automatica*, 32(5):791–792, 1996.

132. T.-Y. Kuc, J. S. Lee, and K. Nam. An iterative learning control theory for a class of nonlinear dynamic systems. *Automatica*, 28(6):1215–1221, 1992.

133. T.-Y. Kuc, J. S. Lee, and B. H. Park. An adaptive hybrid force and position learning control of robot manipulators. In *Proc. of the IEEE Conference on System, Man, and Cybernetics*, pages 2057–2062, San Antonio, TX, USA, Oct. 1994.

134. T.-Y. Kuc, J. S. Lee, and B.-H. Park. Tuning convergence rate of a robust learning controller for robot manipulators. In *Proceedings of the 1995 IEEE Conference on Decision and Control*, volume 2, pages 1714–1719, New Orleans, LA, 1995.

135. J. E. Kurek. Stability of nonlinear parameter-varying digital 2-D systems. *IEEE Trans. of Automat. Contr.*, 40(8):1428–1432, 1986.

136. J. E. Kurek and M. B. Zaremba. Iterative learning control systhesis based on 2-D system theory. *IEEE Trans. of Automatic Control*, 38(1):121–125, 1993.

137. W. C. Lai and P. A. Cook. A discrete-time universal regulator. *Int. J. of Control*, 62(1):17–32, 1995.

138. H. S. Lee and Z. Bien. Memory-based learning control for repetitive motion of robot manipulator. In *Proc. of the Asian Control Conference*, pages 549–552, Tokyo, Japan, Jul. 1994.

139. H.-S. Lee and Z. Bien. Study on robustness of iterative learning control with non-zero initial error. *Int. J. of Control*, 64(3):345–359, 1996.

140. H. S. Lee and Z. Bien. A note on convergence property of iterative learning controller with respect to sup norm. *Automatica*, 33(8):1591–1593, Aug. 1997.

141. H. S. Lee and Z. Bien. Robustness and convergence of a PD-type iterative learning controller. In *Proceedings of the 2nd Asian Control Conference*, Seoul, Korea, July 1997.

142. J.-W. Lee, H.-S. Lee, and Z. Bien. Iterative learning control with feedback using Fourier series with application to robot trajectory tracking. *Robotica*, 11:291–298, 1993.

143. K. H. Lee and Z. Bien. Initial condition problem of learning control. *IEE Proceedings, Part-D, Control Theory and Applications*, 138(6):525–528, 1991.

144. K. H. Lee and Z. Bien. Application of iterative learning controller with feedback loop to the focusing control system of the MODD. In *Proc. of the Asian Control Conference*, pages 243–246, Tokyo, Japan, Jul. 1994.

145. K. S. Lee, S. H. Bang, and K. S. Chang. Feedback-assisted iterative learning control based on an inverse process model. *J. of Process Control*, 4(2):77–89, 1994.

146. K. S. Lee and J. H. Lee. Constrained model-based predictive control combined with iterative learning for batch or repetitive processes. In *Proceedings of the 2nd Asian Control Conference*, Seoul, Korea, July 1997.

147. G. Lee-Glauser, J. N. Juang, and R. W. Longman. Comparison and combination of learning controllers: computational enhancement and experiments. *Journal of Guidance, Control and Dynamics*, 19(5):1116–1123, 1996.

148. A. Leva. PID autotuning algorithm based on relay feedback. *IEE Proc. Pt. D*, 140(5):328–338, 1993.

149. Z.-H. Li, T.-Y. Chai, C. Wen, and C.-B. Soh. Robust output tracking for nonlinear uncertain systems. *Systems and Control Letters*, 25:53–61, 1995.

150. R. F. Lieske and A. M. Mackenzie. Determination of aerodynamic drag from radar data. In *Aberdeen Proving Ground Technical Report*, USA, 1972.

151. H. Lin, L. Wang, and G. Dai. Initial state problem in iterative learning control. In *Proc. of the First Chinese World Congress on Intelligent Control and Intelligent Automation (CWC ICIA '93)*, pages 2269–2273(in Chinese), Beijing, China, Aug. 1993.

152. D. J. Linse and R. F. Stengel. Identification of aerodynamic coefficients using computational neural networks. *Journal of Guidance, Control, and Dynamics*, 16(6):1018–1025, 1994.

153. R. W. Longman, M. Q. Phan, and J. Juang. An overview of a sequence of research developments in learning and repetitive control. In *Proceedings of the First International Conference on Motion and Vibration Control*, Yokohama, Japan, September 1992.

154. R. W. Longman and Y. Wang. Phase cancellation learning control using FFT weighted frequency response identification. *Advances in the Astronautical Sciences*, 93:85–101, 1996.

155. R. W. Longman. Designing Iterative Learning and Repetitive Controllers. In Z. Bien and J.-X. Xu, editors., *Iterative Learning Control - Analysis, Design, Integration and Applications*. Kluwer Academic Publishers, pages 107-146, 1998.

156. A. De Luca, G. Paesano, and G. Ulivi. A frequency-domain approach to learning control: implementation for a robot manipulator. *IEEE Trans. on Industrial Electronics*, 39(1):1–10, 1992.

157. A. De Luca and S. Panzieri. An iterative scheme for learning gravity compensation in flexible robot arms. *Automatica*, 30(6):993–1002, 1994.

158. P. Lucibello. Learning control of linear systems. In *Proc. of American Control Conference*, pages 1888–1892, 1992.

159. P. Lucibello. On the role of high gain feedback in P-type learning control of robots. In *Proc. of the 32nd IEEE Conf. on Decision and Control*, pages 2149–2152, San Antonio, Texas, USA, Dec. 1993.

160. P. Lucibello. Robots repositioning by learning. In *Proc. of the 32nd IEEE Conf. on Decision and Control*, pages 2660–2661, San Antonio, Texas, USA, Dec. 1993.

161. P. Lucibello. Repositioning control of robotic arm by learning. *IEEE Trans. of Automatic Control*, 39(8):1690–1694, Aug. 1994.

162. P. Lucibello. State steering by learning for a class of nonlinear control systems. *Automatica*, 30(9):1463–1468, Sept. 1994.

163. P. Lucibello and S. Panzieri. Cyclic control of linear systems: theory and experimental implementation on a flexible arm. In *Proc. of the 33nd IEEE Conf. on Decision and Control*, pages 369–372, Lake Buena Vista, FL, USA, Dec. 1994.

164. P. Lucibello and S. Panzieri. Experiments on repositioning control of robots by learning. In *Proc. of the Int. Conf. on Robot and Automation*, pages 2910–2914, 1994.

165. P. Lucibello. Output zeroing with internal stability by learning. *Automatica*, 31(11):1665–1672, November 1995.

166. T. Manabe and F. Miyazaki. Learning control based on local linearization by using DFT. *Journal of Robotic Systems*, 11(2):129–141, 1994.

167. W. Messner and R. Horowitz. Identification of a nonlinear function in a dynamical system. *ASME: J. of Dynamic Systems, Measurement, and Control*, 115:587–591, 1993.

168. W. Messner, R. Horowitz, W.-W. Kao, and M. Boals. A new adaptive learning rule. *IEEE Trans. of Automatic Control*, 36(2):188–197, 1991.

169. W. Messner, W.-W. Kao, R. Horowitz, and M. Boals. A new adaptive learning rule. In *Proc. of the 1990 Int. Conf. on Robotics and Automation*, pages 1522–1527, 1990.

170. R. H. Middleton, G. C. Goodwin, and R. W. Longman. A method for improving the dynamic accuracy of a robot performing a repetitive task. *Int. J. of Robotics Research, [See also, University of Newcastle, Newcastle, Australia, Delpartment of electrical Engineering Technical Report EE8546, 1985]*, 8(5):67–74, Oct. 1989.

171. J.-H. Moon, T.-Y. Doh, , and M. J. Chung. A robust approach to iterative learning control design for uncertain systems. *Automatica*, 34(8):1001–1004, 1998.

172. K. L. Moore. *Iterative learning control for deterministic systems*. Advances in Industrial Control. Springer-Verlag, 1993.

173. K. L. Moore. Iterative learning control - an expository overview. *Applied & Computational Controls, Signal Processing, and Circuits*, 1(1):425–488. (e-copy: http://www.engineering.usu.edu/ece/faculty/moorek/survey.zip).

174. K. L. Moore, M. Dahleh, and S. P. Bhattacharyya. Iterative learning control: a survey and new results. *J. of Robotic Systems*, 9(5):563–594, 1992.

175. K. L. Moore. *Design Techniques for Transient Response Control*. PhD thesis, Texas A&M University, College Station, Texas, 1989.

176. T. Naniwa and S. Arimoto. Learning control for robottasks under geometric endpoint constraints. *IEEE Trans. of Robotics and Automation*, 11(3):432–440, 1991.

177. T. Naniwa, S. Arimoto, and L. L. Whitcomb. Learning control for robot tasks under geometric constraints. In *Proc. of the 1994 IEEE Int. Conf. on Robotics and Automation*, pages 2921–2927, 1994.

178. R. H. Nathan. Control strategies in FNS systems for the upper extremities. *Critical Reviews in Biomedical Engineering*, 21(6):485–568, 1993.

179. Z. Novakovic. *The principle of self-support in control systems*, volume 8 of *Studies in Automation and Control*. Elsevier Science Publisher B.V., Amsterdam, 1992.

180. Z. Novakovic. Robust tracking control for robots with bounded input. *ASME: J. of Dynamic Systems, Measurement, and Control*, 114:315–319, 1992.

181. S. R. Oh, Z. Bien, and I. H. Suh. An iterative learning control method with application for the robot manipulator. *IEEE J. of Robotics and Automation*, 4(5):508–514, 1988.

182. G. Oriolo, S. Panzieri, and G. Ulivi. Cyclic learning control of chain-form systems with applications to car-like robots. In *Proc. of 13-th Triennial World Congress of IFAC*, pages 187–192, San Francisco, USA, 1996.

183. D. H. Owens. Stability of linear multipass processes. *Proc. IEE*, 124(11):1079–1082, 1977.

184. D. H. Owens. Iterative learning control - convergence using high gain feedback. In *Proc. of the 31st Conf. on Decision and Control*, pages 2545–2546, Tucson, Arizona, USA, Dec. 1992.

185. D. H. Owens. Universal iterative learning control using adaptive high-gain feedback. *Int. J. of Adaptive Control and Signal Processing*, 7:383–388, 1993.

186. D. H. Owens, N. Amann, and E. Rogers. Iterative learning control - an overview of recent algorithms. *Applied Mathematics and Computer Science*, 5(3):425–438, 1995.

187. M. Pandit and K. Buchheit. Iterativ lernende regelung zyklischer produktionsprozesse. *Automatisierungstechnik*, 44(1):21–31, 1996.

188. K.-H. Park, Z. Bien, and D.-H. Hwang. Design of an iterative learning controller for a class of linear dynamic systems with time- delay. In *Proceedings of the 2nd Asian Control Conference*, Seoul, Korea, July 1997.

189. D. Peyton, H. Kinoshita, G. Q. Lo, and D. L. Kwong. Systems oriented survey of noncontact temperature measurement techniques for rapid thermal processing. In *Rapid Thermal and Related Processing Techniques, Vol.1393*, pages 259–308, October 1990.

190. M. Phan and R. W. Longman. A mathematical theory of learning control for linear discrete multivariable systems. In *Proc. of the AIAA/AAS Astrodynamics Specialist Conference*, Minneapolis, Minnesota, USA, August 1988.

191. M. Phan and R. W. Longman. Learning control for trajectory tracking using basis functions. In *Proc. of the IEEE Conference on Decision and Control*, pages 2490–2492, Kobe, Japan, Dec. 1996.

192. M. Q. Phan and J. N. Juang. Designs of learning controllers based on an auto-regressive representation of a linear system. *AIAA Journal of Guidance, Control, and Dynamics*, 19(2):355–362, 1996.

193. A. N. Poo, K. B. Lim, and Y. X. Ma. Application of discrete learning control to a robotic manipulator. *Robotics and Computer-Integrated Manufacturing*, 12(1):55–64, March 1996.

194. B. Porter and S. S. Mohamed. Iterative learning control of partially irregular multivariable plants with initial state shifting. *International Journal of Systems Science*, 22(2):229–235, 1991.

195. B. Porter and S. S. Mohamed. Iterative learning control of partially irregular multivariable plants with initial impulsive action. *International Journal of Systems Science*, 22(3):447–454, 1991.

196. Z. Qu and D. M. Dawson. Lyapunov direct design of robust control for electrical-mechanical systems of cascaded nonlinear uncertain systems. *Trans. of ASME: J. of Dynamic Systems, Measurement, and Control*, 117:54–62, 1995.

197. Z. Qu, J. Dorsey, D. M. Dawson, and R. W. Johnson. A new learning control scheme for robots. In *Proc. of the 1991 IEEE Conf. on Robotics and Automation*, pages 1463–1468, Sacramento, CA, USA, Apr. 1991.

198. Z. Qu and H. Zhuang. Non-linear learning control of robot manipulators without requiring acceleration measurement. *Int. J. of Adaptive Control and Signal Processing*, 7:77–90, 1993.

199. R. D. Reinke, D. Pratzel-Wolters, and D.H. Owens. Algorithms for computing the solutions of 1D and 2D Lyapunov equations for discrete linear repetitive processes. In *Proc. of European Control Conference*, pages 2220–2225, Rome, Italy, Sept. 1995.

200. X. Ren and W. Gao. On the initial conditions in learning control. In *Proc. of the IEEE Int. Sym. on Industry Electronics*, pages 182–185, Xi'an, China, 1992.

201. P. D. Roberts. Unit memory repetitive process aspects of iterative optimal control. In *Proc. of the 33rd Conf. on Decision and Control*, pages 1394–1399, Lake Buena Vista, FL, USA, Dec. 1994.

202. E. Rogers and D. H. Owens. Controller design for industrial multipass processes. In *Proc. of the 3rd European Conference for Mathematics in Industry*, pages 495–502, Glasgow, UK, 1988.

203. E. Rogers and D. H. Owens. Stability and state feedback control of differential unit memory linear multipass processes. In *Proc. of American Control Conference*, pages 51–51, 1988.

204. E. Rogers and D. H. Owens. 2D transfer-functions and stability tests for differential unit memory linear multipass processes. *International Journal of Control*, 50(2):651–666, 1989.

205. E. Rogers and D. H. Owens. Output feedback control of linear multipass processes. In *Proc. of American Control Conference*, pages 318–319, Pittsburgh, PA, USA, 1989.

206. E. Rogers and D. H. Owens. Stability analysis for discrete linear multipass processes with non-unit memory. *IMA Journal of Mathematical Control and Information*, 6(4):399–409, 1989.

207. E. Rogers and D. H. Owens. *Stability Analysis for Linear Repetitive Processes.* Springer-Verlag, Berlin, 1992.

208. E. Rogers and D. H. Owens. Output-feedback control of discrete linear repetitive processes. *IMA Journal of Mathematical Control and Information*, 10:177–193, 1993.

209. E. Rogers and D. H. Owens. Error actuated feedback control theory for differential linear repetitive processes. *International Journal of Control*, pages –, 1995.

210. E. Rogers and D. H. Owens. Stability of linear repetitive processes - a delay-differential systems interpretation. *IMA Journal of Mathematical Control and Information*, 12:69–90, 1995.

211. C. L. Roh, M. N. Lee, and M. J. Chung. ILC for non-minimum phase systems. *International Journal of Systems Science*, 27(4):419–424, April 1996.

212. F. Roozeboom and N. Parekh. Rapid thermal processing systems: A review with emphasis on temperature control. *J. Vac. Sci. Technol. B*, 8(6):1249–1259, November 1990.

213. Y. S. Ryu and R. W. Longman. Use of anti-reset windup in integral control based learning and repetitive control. In *Proc. of the IEEE Conference on System, Man, and Cybernetics*, pages 2617–2622, San Antonio, TX, USA, Oct. 1994.

214. S. S. Saab. *Learning control: convergence and robustness*. PhD thesis, University of Pittsburgh, 1992.

215. S. S. Saab. On the P-type learning control. *IEEE Trans. of Automatic Control*, 39(11):2298–2302, 1994.

216. S. S. Saab. A discrete-time learning control algorithm for a class of linear time-invariant systems. *IEEE Trans. of Automatic Control*, 40(6):1138–1141, 1995.

217. S. S. Saab. Discrete-time learning control algorithm for a class of nonlinear systems. In *Proc. of American Control Conference*, pages 2739–2743, Seattle, Washington, USA, June 1995.

218. S. S. Saab, W. G. Vogt, and M. H. Mickle. Robustness and convergence of P-type learning control. In *Proc. of American Control Conference*, pages 36–38, San Francisco, CA, USA, Jun. 1993.

219. N. Sadegh. A discrete-time MIMO repetitive controller. In *Proc. of American Control Conference*, pages 2671–2675, 1992.

220. N. Sadegh. Synthesis of a stable discrete-time repetitive controller for MIMO systems. *Journal of Dynamic Systems, Measurement, and Control*, 117:92–98, March 1995.

221. N. Sadegh, R. Horowitz, and W.-W. Kao. A unified approach to the design of adaptive and repetitive controllers for robotic manipulators. *Journal of Dynamic Systems, Measurement, and Control*, 112:618–629, December 1990.

222. W. E. Singhose and N. C. Singer. Effects of input shaping on two-dimentional trajectory following. *IEEE Trans. on Robotics and Automation*, 12(6):881–887, Dec. 1996.

223. R. B. Stein, P. H. Peckham, and D. B. Popovic. *Neural prosthesis: replacing motor function after disease or disability*. New York: Oxford University Press, 1992.

224. Y. Stepanenko and X. Yang. Stabilization analysis of discrete nonlinear systems. *Int. J. of Control*, 62(6):1313–1326, 1995.

225. T. Sugie and T. Ono. An iterative learning control law for dynamical systems. *Automatica*, 27(4):729–732, 1991.

226. D. Sun, X. Shi, and Y. Liu. Adaptive learning control for cooperation of two robots manipulating a rigid object with model uncertainties. *Robotica*, 14(4):365–373, July-August 1996.

227. M. Sun, D. Ye, and B. Wan. A PI-type iterative learning scheme for nonlinear dynamical systems. In *Proc. of the First Chinese World Congress on Intelligent Control and Intelligent Automation (CWC ICIA '93)*, pages 2279–2283, Beijing, China, Aug. 1993.

228. M. Sun, B. Huang, X. Zhang, and Y. Chen. Robust convergence of the D-type learning controller. In *Proc. of The Second Chinese World Congress on Intelligent Control and Intelligent Automation (CWC ICIA '97)*, Xi'an Jiaotong University Press., pages 678–83, Xi'an, P.R. China, June 23-27 1997.

229. M. Sun, B. Huang, X. Zhang, Y. Chen, and J.-X. Xu. Selective learning with a forgetting factor for trajectory tracking of uncertain nonlinear systems. In *Proceedings of the 2nd Asian Control Conference*, Seoul, Korea, July 1997.

230. N. Sureshbabu and W. J. Righ. On output regulation for discrete-time nonlinear systems. *Control Theory and Technology*, 10(2):281–195, 1994.

231. N. Sureshbabu and W. J. Righ. On output regulation for discrete-time nonlinear systems. In *Proc. of American Control Conf.*, pages 4226–4230, Seattle, Washington, June 1995.

232. T. Suzuki, M. Yasue, S. Okuma, and Y. Uchikawa. Discrete-time learning control for robotic manipulators. *Advanced Robotics*, 9(1):1–14, Jan. 1995.

233. K. K. Tan. *Towards autonomous process control with relay-based feedback identification*. PhD thesis, National University of Singapore, 10 Kent Ridge Crescent, Dec. 1994.

234. K. K. Tan, H. F. Dou, Y. Q. Chen and T. H. Lee. High precision linear motor control via relay-tuned iterative learning based on zero-phase filtering . *IEEE Transcation on Control System Technology*. (submitted) 1999.

235. M. Togai and O. Yamano. Analysis design of an optimal learning control scheme for industrial robots: a discrete system approach. In *Proc. of the 24th Conf. on Decision and Control*, pages 1399–1404, Ft. Lauderdale, FL., Dec. 1985.

236. T.-C. Tsao and Y.-X. Qian. An adaptive repetitive control scheme for tracking periodic signals with unknown period. In *Proc. of American Control Conference*, pages 1736–1740, San Francisco, CA, USA, Jun. 1993.

237. T.-C. Tsao and M. Tomizuka. Robust adaptive and repetitive digital tracking control and application to a hydraulic servo for noncircular machining. *Journal of Dynamic Systems, Measurement, and Control*, 116:24–32, March 1994.

238. S. K. Tso and L. Y. X. Ma. Discrete learning control for robots: strategy, convergence and robustness. *International Journal of Control*, 57(2):273–291, 1993.

239. S. K Tso and L. Y. X. Ma. A self-contained iterative learning controller for feedback control of linear systems. In *Proc. of the Asian Control Conference*, pages 545–548, Tokyo, Japan, Jul. 1994.

240. S. K. Tso and Y. X. Ma. Cartesian-based learning control for robots in discrete-time formulation. *IEEE Transactions on Systems, Man, and Cybernetics*, 22, September 1992.

241. S. K Tso, C. S. Tam, and L. Y. X. Ma. Position and velocity feedback for iterative learning control of robot manipulators. In *Proc. of the Asian Control Conference*, pages 235–238, Tokyo, Japan, Jul. 1994.

242. Y. Uno, M. Kawato, and R. Suzuki. Formation and control of optimal trajectory in human multi-joint arm movement. *Biological Cybernetics*, 61:89–101, 1989.

243. D. Wang. An anticipatory iterative learning control scheme: theory and experiments. In K. L. Moore, editor, *Proc. of the First Int. Workshop on Iterative Learning Control*, pages 79–80, Hyatt Regency, Tampa, FL, USA, Dec. 1998.

244. D. Wang, Y. C. Soh, and C. C. Cheah. Robust motion and force control of constrained manipulators by learning. *Automatica*, 31(2):257–262, 1995.

245. C. Wen and D. J. Hill. Global boundedness of discrete-time adaptive control just using estimator projection. *Automatica*, 28(6):1143–1157, 1992.

246. W. E. Jr. Williamson. Instrument modeling for aerodynamic coefficient identification from flight test data. *AIAA Journal of Guidance and Control*, 3(3):225–279, 1980.

247. J. - X. Xu and Z. Bien, Frontiers in Iterative Learning Control. In Z. Bien and J.-X. Xu, editors., *Iterative Learning Control - Analysis, Design, Integration and Applications*. Kluwer Academic Publishers, pages 9-36, 1998.

248. J.-X. Xu. Direct learning of control input profiles with different time scales. In *Proceedings of the 35th IEEE Conference on Decision and Control*, Kobe, Japan, December 1996.

249. J.-X. Xu, Y. Chen, T. H. Lee, and S. Yamamoto. Terminal iterative learning control with an application to RTPCVD thickness control. *Automatica*, 35(9), Sept. 1999.

250. J.-X. Xu and Z. Qu. Robust learning control for a class of non-linear systems. In *Proceedings of the 35th IEEE Conference on Decision and Control*, Kobe, Japan, December 1996.

251. J.-X. Xu, X.-W. Wang, and T. H. Lee. Analysis of continuous iterative learning control systems using current cycle feedback. In *Proc. of American Control Conference*, pages 4221–4225, Seattle, Washington, USA, June 1995.

252. J.-X. Xu, Y. Dote, X. Wang, and B. Shun. On the instability of iterative learning control due to sampling delay. In *Proceedings of the 1995 IEEE IECON 31st International Conference on Industrial Electronics, Control, and Instrumentation*, volume 1, pages 150–155, Orlando, FL, November 1995.

253. J.-X. Xu and Y. Song. Direct learning control of high-order systems for trajectories with different time scales. In *Proceedings of the 2nd Asian Control Conference*, Seoul, Korea, July 1997.

254. J.-X. Xu and T. Zhu. Direct learning control of trajectory tracking with different time and magnitude scales for a class of nonlinear uncertain systems. In *Proceedings of the 2nd Asian Control Conference*, Seoul, Korea, July 1997.

255. S. Yamamoto and I. Hashimoto. Resent status and future needs: The view from Japanese industry. In Arkun and Ray, editors, *Proceedings of the fourth International Conference on Chemical Process Control*, Texas, 1991. Chemical Process Control – CPCIV.

256. E. Zafiriou, R.A. Adomaitis, and G. Gattu. An approach to run-to-run control for rapid thermal processing. In *Proc. of American Control Conf.*, pages 1286–1288, Seattle, WA, USA, 1995.

257. N. Zeng and X. Ying. Iterative learning control algorithm for nonlinear dynamical systems. *Acta Automatica Sinica*, 18(2):168–176 (in Chinese), Mar. 1992.

258. B. S. Zhang and J. R. Leigh. Predictive time-sequence iterative learning control with application to a fermentation process. In *Proc. of the 2nd IEEE Conf. on Control Applications*, pages 439–442, Vancouver, B.C., Canada, Sept. 1993.

259. P. Zhang, Y. Sankai, and M. Ohta. Hybrid adaptive learning control of nonlinear system. In *Proc. of American Control Conference*, pages 2744–2748, Seattle, Washington, June 1995.

Index

Lecture Notes in Control and Information Sciences

Edited by M. Thoma

1993–1999 Published Titles:

Vol. 203: Popkov, Y.S.
Macrosystems Theory and its Applications:
Equilibrium Models
344 pp. 1995 [3-540-19955-1]

Vol. 204: Takahashi, S.; Takahara, Y.
Logical Approach to Systems Theory
192 pp. 1995 [3-540-19956-X]

Vol. 205: Kotta, U.
Inversion Method in the Discrete-time
Nonlinear Control Systems Synthesis
Problems
168 pp. 1995 [3-540-19966-7]

Vol. 206: Aganovic, Z.; Gajic, Z.
Linear Optimal Control of Bilinear Systems
with Applications to Singular Perturbations
and Weak Coupling
133 pp. 1995 [3-540-19976-4]

Vol. 207: Gabasov, R.; Kirillova, F.M.;
Prischepova, S.V.
Optimal Feedback Control
224 pp. 1995 [3-540-19991-8]

Vol. 208: Khalil, H.K.; Chow, J.H.;
Ioannou, P.A. (Eds)
Proceedings of Workshop on Advances
inControl and its Applications
300 pp. 1995 [3-540-19993-4]

Vol. 209: Foias, C.; Özbay, H.;
Tannenbaum, A.
Robust Control of Infinite Dimensional
Systems: Frequency Domain Methods
230 pp. 1995 [3-540-19994-2]

Vol. 210: De Wilde, P.
Neural Network Models: An Analysis
164 pp. 1996 [3-540-19995-0]

Vol. 211: Gawronski, W.
Balanced Control of Flexible Structures
280 pp. 1996 [3-540-76017-2]

Vol. 212: Sanchez, A.
Formal Specification and Synthesis of
Procedural Controllers for Process Systems
248 pp. 1996 [3-540-76021-0]

Vol. 213: Patra, A.; Rao, G.P.
General Hybrid Orthogonal Functions and
their Applications in Systems and Control
144 pp. 1996 [3-540-76039-3]

Vol. 214: Yin, G.; Zhang, Q. (Eds)
Recent Advances in Control and Optimization
of Manufacturing Systems
240 pp. 1996 [3-540-76055-5]

Vol. 215: Bonivento, C.; Marro, G.;
Zanasi, R. (Eds)
Colloquium on Automatic Control
240 pp. 1996 [3-540-76060-1]

Vol. 216: Kulhavý, R.
Recursive Nonlinear Estimation: A Geometric
Approach
244 pp. 1996 [3-540-76063-6]

Vol. 217: Garofalo, F.; Glielmo, L. (Eds)
Robust Control via Variable Structure and
Lyapunov Techniques
336 pp. 1996 [3-540-76067-9]

Vol. 218: van der Schaft, A.
L_2 Gain and Passivity Techniques in Nonlinear
Control
176 pp. 1996 [3-540-76074-1]

Vol. 219: Berger, M.-O.; Deriche, R.;
Herlin, I.; Jaffré, J.; Morel, J.-M. (Eds)
ICAOS '96: 12th International Conference on
Analysis and Optimization of Systems -
Images, Wavelets and PDEs:
Paris, June 26-28 1996
378 pp. 1996 [3-540-76076-8]

Vol. 220: Brogliato, B.
Nonsmooth Impact Mechanics: Models,
Dynamics and Control
420 pp. 1996 [3-540-76079-2]

Vol. 221: Kelkar, A.; Joshi, S.
Control of Nonlinear Multibody Flexible Space
Structures
160 pp. 1996 [3-540-76093-8]

Vol. 222: Morse, A.S.
Control Using Logic-Based Switching
288 pp. 1997 [3-540-76097-0]

Vol. 223: Khatib, O.; Salisbury, J.K.
Experimental Robotics IV: The 4th
International Symposium, Stanford, California,
June 30 - July 2, 1995
596 pp. 1997 [3-540-76133-0]

Vol. 224: Magni, J.-F.; Bennani, S.;
Terlouw, J. (Eds)
Robust Flight Control: A Design Challenge
664 pp. 1997 [3-540-76151-9]

Vol. 225: Poznyak, A.S.; Najim, K.
Learning Automata and Stochastic
Optimization
219 pp. 1997 [3-540-76154-3]

Vol. 226: Cooperman, G.; Michler, G.;
Vinck, H. (Eds)
Workshop on High Performance Computing
and Gigabit Local Area Networks
248 pp. 1997 [3-540-76169-1]

Vol. 227: Tarbouriech, S.; Garcia, G. (Eds)
Control of Uncertain Systems with Bounded
Inputs
203 pp. 1997 [3-540-76183-7]

Vol. 228: Dugard, L.; Verriest, E.I. (Eds)
Stability and Control of Time-delay Systems
344 pp. 1998 [3-540-76193-4]

Vol. 229: Laumond, J.-P. (Ed.)
Robot Motion Planning and Control
360 pp. 1998 [3-540-76219-1]

Vol. 230: Siciliano, B.; Valavanis, K.P. (Eds)
Control Problems in Robotics and Automation
328 pp. 1998 [3-540-76220-5]

Vol. 231: Emel'yanov, S.V.; Burovoi, I.A.;
Levada, F.Yu.
Control of Indefinite Nonlinear Dynamic
Systems
196 pp. 1998 [3-540-76245-0]

Vol. 232: Casals, A.; de Almeida, A.T. (Eds)
Experimental Robotics V: The Fifth
International Symposium Barcelona,
Catalonia, June 15-18, 1997
190 pp. 1998 [3-540-76218-3]

Vol. 233: Chiacchio, P.; Chiaverini, S. (Eds)
Complex Robotic Systems
189 pp. 1998 [3-540-76265-5]

Vol. 234: Arena, P.; Fortuna, L.; Muscato, G.;
Xibilia, M.G.
Neural Networks in Multidimensional
Domains: Fundamentals and New Trends in
Modelling and Control
179 pp. 1998 [1-85233-006-6]

Vol. 235: Chen, B.M.
H∞ Control and Its Applications
361 pp. 1998 [1-85233-026-0]

Vol. 236: de Almeida, A.T.; Khatib, O. (Eds)
Autonomous Robotic Systems
283 pp. 1998 [1-85233-036-8]

Vol. 237: Kreigman, D.J.; Hagar, G.D.;
Morse, A.S. (Eds)
The Confluence of Vision and Control
304 pp. 1998 [1-85233-025-2]

Vol. 238: Elia, N. ; Dahleh, M.A.
Computational Methods for Controller Design
200 pp. 1998 [1-85233-075-9]

Vol. 239: Wang, Q.G.; Lee, T.H.; Tan, K.K.
Finite Spectrum Assignment for Time-Delay
Systems
200 pp. 1998 [1-85233-065-1]

Vol. 240: Lin, Z.
Low Gain Feedback
376 pp. 1999 [1-85233-081-3]

Vol. 241: Yamamoto, Y.; Hara S.
Learning, Control and Hybrid Systems
472 pp. 1999 [1-85233-076-7]

Vol. 242: Conte, G.; Moog, C.H.; Perdon A.M.
Nonlinear Control Systems
192 pp. 1999 [1-85233-151-8]

Vol. 243: Tzafestas, S.G.; Schmidt, G. (Eds)
Progress in Systems and Robot Analysis and
Control Design
624 pp. 1999 [1-85233-123-2]

Vol. 244: Nijmeijer, H.; Fossen, T.I. (Eds)
New Directions in Nonlinear Observer Design
552pp: 1999 [1-85233-134-8]

Vol. 245: Garulli, A.; Tesi, A.; Vicino, A. (Eds)
Robustness in Identification and Control
448pp: 1999 [1-85233-179-8]

Vol. 246: Aeyels, D.;
Lamnabhi-Lagarrigue, F.; van der Schaft, A. (Eds)
Stability and Stabilization of Nonlinear Systems
408pp: 1999 [1-85233-638-2]

Vol. 247: Young, K.D.; Özgüner, Ü. (Eds)
Variable Structure Systems, Sliding Mode and
Nonlinear Control
400pp: 1999 [1-85233-197-6]